# Robotic Fabrication & Architectural Design

EDITED BY

ROBERT STUART-SMITH

MSD-RAS
Robotics & Autonomous Systems
UNIVERSITY OF PENNSYLVANIA, WEITZMAN SCHOOL OF DESIGN

ORO EDITIONS

*Robotic Fabrication & Architectural Design* presents design research from the University of Pennsylvania Stuart Weitzman School of Design's MSD-RAS program.

At present, architectural design and construction approaches are unable to meet immediate and projected societal needs in productivity, affordability, and sustainability or to adequately engage with the diverse conditions found in our built environment. The Master of Science in Design: Robotics and Autonomous Systems degree (MSD-RAS), seeks to address these challenges through bespoke design solutions that are integral to a critical and creative approach to production. Implied in the term "RAS", the program seeks to harness the potential of AI and robotic systems to work more adaptively than automation affords. Primarily operating through the development of robotically fabricated prototypes, projects that incorporate custom approaches to generative computational design, machine learning, robot tooling, real-time adaptive robot programming, sensor feedback, material and manufacturing processes or human-in-the-loop activities are presented. Serving as a graphical reflection on the first three years of the program, the book showcases student research projects alongside interviews with graduates, and other insights into the exciting career paths graduates have pursued after their studies. Essays from the program's faculty dive deeper into several core topics such as the MSD-RAS's approach to design research, critical engagement with industrial manufacturing processes, and the integration of semi-autonomous workflows in design and production. Also discussed is the program's unique integrated approach to coursework and why it is inducive to the creation of novel collaborative work that expands design agency into unchartered territories and careers.

***Including Conributions by:*** Jeffrey Anderson, Billie Faircloth, Nathan King, Andrew Saunders, Ezio Blasetti, Patrick Danahy, Jose Luis Garcia del Castillo Lopez, and Evangelos Kotsioris.

*Robert Stuart-Smith* is the Director of the MSD-RAS degree program, Assistant Professor of Architecture, and Affiliate Faculty in Engineering's GRASP Lab at the University of Pennsylvania. He leads the Autonomous Manufacturing Lab in Penn's Department of Architecture and University College London's Department of Computer Science, managing over $5 million in research and collaborating with industry partners such as Cemex, Skanska, Mace, Burohappold, and Ultimaker. Stuart-Smith's research intersects design, computation, robotic fabrication, and collective robotic construction. His Aerial Additive Manufacturing research, published in Nature, demonstrates the first in-flight additive manufacturing by cooperating drones. Integrating robotic manufacturing with architectural design, Stuart-Smith addresses the environmental and economic costs of production while enhancing their cultural and aesthetic impacts. He co-directs Robert Stuart-Smith Design and was a co-founder of Kokkugia. Author of the book *Behavioural Production*, he has published in journals including Nature, Science Robotics, and AD Architectural Design. Included in the permanent collection at Frac Centre-Val de Loire, his work has been exhibited at Venice, Tallinn, Beijing, and Prague Architecture Biennales. He has taught at AA, WashU, RMIT, U.Innsbruck, and lectured at institutions such as ETHZ, U.Stuttgart, MIT, CCA, Sci-Arc, AA, Angewandte, Strelka Institute, and Tsinghua. His work has been featured by BBC Click, New Scientist, Smithsonian, Architizer, France 3, and others.

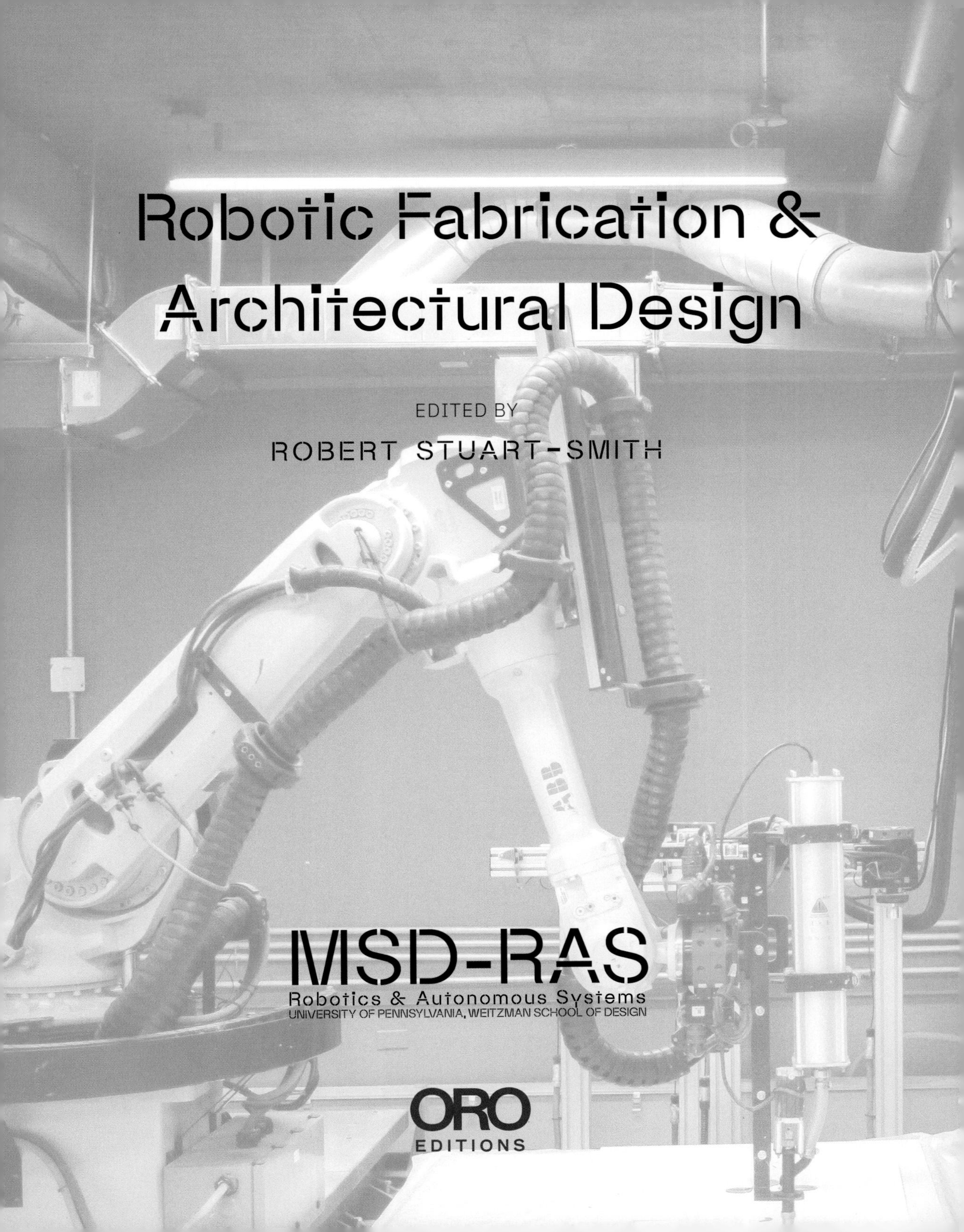

# Robotic Fabrication & Architectural Design

EDITED BY

ROBERT STUART-SMITH

MSD-RAS

Robotics & Autonomous Systems

UNIVERSITY OF PENNSYLVANIA, WEITZMAN SCHOOL OF DESIGN

ORO

EDITIONS

# CONTENTS

# ACKNOWLEDGEMENTS

This book primarily showcases student projects from the first three years of the Weitzman School of Design's new post-professional Masters of Science in Design: Robotics and Autonomous Systems program (MSD-RAS) at the University of Pennsylvania. As will be apparent to the reader, the work is ambitious and full of exciting and real potential. Students contributing to each project are acknowledged adjacent to their work within the book.

This work could not have been achieved without the specialist knowledge, curricula, and commitment of outstanding MSD-RAS faculty including Andrew Saunders, Evangelos Kotsioris, Ezio Blasetti, Billie Faircloth, Nathan King, Jeffrey Anderson, Jose Luis Garcia del Castillo Lopez, Alicia Nahmad Vazquez, Patrick Danahy, Emek Erdolu, and Teaching Assistants Patrick Danahy, David Forero, Hadi El Kebib, Sophia O'Neil, Riley Studebaker, Claire Moriarty, Matthew White, Caleb Ehly, and Yujie Li.

Students undertook most of their work within the Weitzman School of Design's Advanced Research and Innovation (ARI) Robotics Lab, supported by ARI Robotics Lab staff. Special thanks to ARI Lab managers Nicholas Sideropoulos and Shunta Moriuchi and former managers Sophia O'Neil, Riley Studebaker, and David Forero together with part-time employees who supported work in the lab through operational, maintenance, research and development activities. Thank you also to Weitzman's Senior Director for Operations, Planning, Design and Construction – Karl Wellman who has been an awesome partner and collaborator with me in jointly coordinating this staff and facility.

The foundations of the MSD-RAS program were laid by the former two-term Chair of Architecture Winka Dubbeldam and Standing Faculty in the Department of Architecture, who had the insight and initiative to create the potential for an MSD program and a facility centred around robotic fabrication. I am honored to have been invited to join the faculty and tasked with developing the new ARI Robotics Lab facility and MSD program in robotics. As Director of the MSD-RAS program, I had a unique opportunity to propose a new program, its curricula, and the hiring of faculty in parallel to developing a new robotics lab space and its equipment. This was a momentous task that would not have been possible without the active and instrumental support of Winka Dubbeldam; Weitzman School of Design's Dean Fritz Steiner; Vice Dean Leslie Hurtig; Associate Chair Andrew Saunders; Senior Director for Operations and Planning Karl Wellman; Assistant Dean of Development and Alumni Relations Jeffrey Snyder; Director of Finance Chris Cataldo; Senior Director of Communications Michael Grant; the Department of Architecture Faculty; the Weitzman administration team; and architects for the ARI Robotics Lab renovation – Erdy McHenry Architects. The newly appointed Chair of Architecture, Rossana Hu, has also already provided thoughtful inspiration and direction for the future growth and impact of the program.

The MSD-RAS program's capacity to do frontier work in robotic fabrication is quite obviously dependent on access to advanced facilities and equipment that can support such work. In this context, the completion of the Weitzman School's ARI Robotics Lab was essential to the running of the MSD-RAS program. Much gratitude is extended to the Weitzman School of Design and its Board of Advisors who contributed and raised funds to support a renovation of the ARI Robotics Lab space and the purchase of equipment which enabled the lab to be operational in time for the commencement of the MSD-RAS program in 2020 – in the midst of the Covid-19 global pandemic with great success! A competitively awarded Research Facilities Development Fund (RFDF) from the Vice Provost for Research's office was also instrumental in facilitating the completion of the lab.

Beyond support from the Weitzman School of Design, the University of Pennsylvania Provost's Office, and the Penn community, many in the wider, global academic community and industry have contributed through lectures, jury reviews, round table discussions, and other events, providing insightful feedback on individual projects and the ongoing development of the program. Jurors who participated in each final review are mentioned within the book.

A vast number of people have played a core role in the successful launch and operation of the MSD-RAS program. The program focuses on team-based design research, and the leadership and administrative staff, faculty, curricula, and teaching methods also operate alongside student work to support a larger collaborative venture. This ambitious collective is dedicated to critical and creative design research and aims to positively impact the future of the built environment. The built environment faces several challenges that only collaborative efforts can overcome. The work presented in this book represents an initial step towards achieving this goal, by putting forward design possibilities that are both industry-relevant and speculative. Through this book and its constituent project's published peer-reviewed research papers, the program's work is contributing to globally shared knowledge and helps to build a shared picture of where architectural design, fabrication, and construction can go from here.

*— Robert Stuart-Smith*

*MSD-RAS Program Director, Weitzman School of Design, University of Pennsylvania.*

# ARCHITECTURE 4.0

## *Critical and Creative Approaches to Production for the Fourth Industrial Revolution*

On a warm and otherwise relaxing summer's day this August, *The Guardian* featured an article by Oliver Wainright (Wainwright 2023) that speculated on the emergence of Large Language Model AI systems that are able to generate novel images from a person's text prompts. The article questioned whether such tools might automate so many tasks as to render the architect's role obsolete or vastly shrink the number of architects employed on projects. What was perhaps even more disconcerting was the vast outpouring of commentary from a readership that included many who saw architects adding unnecessary complexity, cost, and delay to otherwise streamlined building processes, in addition to those who believed they could design their own renovations using the latest smartphone AI-based design application (Wainwright 2023).

It is unusual for an independent newspaper like *The Guardian* to lightly raise such popular topics. Given that it is generally believed that AI software is on the cusp of partially or wholly automating several highly-skilled, knowledge-based industries, the article should give reason to pause. The International Monetary Fund just published a report predicting that more than 60% of jobs in the US, UK, and Europe (and 40% globally) will be impacted by AI. At least half of this will negatively affect employment and income prospects, leading to greater inequality (Plumb 2024). This marks a turning point, a technology-driven paradigm shift set to bring massive societal and industry change. Instead of fearing the automation of professional tasks or a reduction in vocational employment opportunities, it would be more astute to ask what the role of the architect should be, or even better,

*What type of built environment does society need and how best could we leverage technological developments to better provide this?*

What agency is needed to achieve such aspirations and what educational models could facilitate this? The field of architecture can only retain or expand its relevance by rising to the challenges of today and tomorrow.

### Systemic Challenges

There are fundamental, systemic problems with the way buildings are designed and constructed that are not able to be resolved through antiquated models of practice and education or through uncritical means of automation. Building supply is already an unmet challenge. At present, international housing supply is insufficient for current needs. There are more than 110 million forcibly displaced people globally without adequate housing (UNHCR 2023). With estimated population growth, there is also a projected additional 2.5 billion people expected to be living in urban centers by 2050 (UN DESA 2014). In just 27 years, current inadequate productivity levels must be substantially improved to meet projected growth. Unfortunately, today's building practices are not up to the task. Globally, the construction industry is the least productive manufacturing industry. In the US, where agriculture and manufacturing sectors

*Previous page) Close-up of additively manufactured and robotically smooshed(manipulated) ceramic façade prototype (see Robotic Prometheus project, spring semester). Below) Slip-cast elements embody a marbling material effect produced using and right) custom robot end-of-arm tool and a robotic roto-molding process (see "These Things Swirl", spring semester).*

expanded productivity by 10-15 times since the 1950s, construction productivity has barely changed (McKinsey Global Institute 2017, 1). In some of the largest economies, such as the US and UK, the skilled construction labor force is also in decline. Quite possibly due to construction work remaining extremely dangerous, with two to three times the fatalities of other manufacturing industries, and between three and six times in emerging economies (International Labour Organisation 2015).

Buildings already consume 40% of global raw materials and 36% of energy annually, producing 39% of global CO2 emissions (International Energy Agency (IEA) 2019). In the US, 30% of total yearly building material is also wasted (Osmani 2011, 209). To address the world's present and projected needs requires a significant increase in productivity, which will inevitably increase construction's large environmental impact, posing an existential threat to our already fragile environment, likely to accelerate climate change and environmentally driven inequality. This lack of productivity, worker safety, and environmental efficiency, together with a host of other concerns such as affordability and quality assurance, collectively indicate that there is substantial opportunity to improve the status quo by increasing industrial methods of production.

Industrial production has historically been associated with a loss of local culture, artisanal craft, and creative diversity in the built environment. In this context, architects such as Anupama Kundoo, Boonserm Premthada, or Anna Heringer provide a humanizing voice. Representative of a globally reemerging concern for local knowledge and culture, these architects seek to reinforce local practices and building methods in their unique architectural work (Harrouk 2023; Heringer n.d.; Kundoo et al. 2020). Kundoo in particular, in recognizing the decline in artisanal work in her home country of India, collaborates with local craftspeople, ensuring local human labor is embodied in her projects, while also producing beautiful bespoke buildings that harness local knowledge and material practices. This work stretches from humanitarian structures such as the Volontariat Home for Homeless Children in Puducherry (Bose 2022) to more elaborate villas, such as her own Wall House that incorporates locally made bricks and unique roof vaults formed from locally produced Terracotta pots (Desai 2016, 314).

Three years after Wall House was partially reconstructed in the 2012 Venice Architecture Biennale (Desai 2016, 318), the world's largest housing market, India was estimated to need an additional 170 million homes within 15 years to house its increasingly urban dwelling population (Betts et al. 2015, 15). Although Kundoo is designing larger developments in her ambitious work for Auroville, neither Kundoo nor other architect's designs offer a systemically scalable approach to high-volume production of multi-story urban construction. To achieve such extreme numbers, India's buildings must be built faster and cheaper, yet globally, housing is increasingly unaffordable and unable to keep up with demand. Despite employing cheaper materials and

working methods than is typically used in other forms of manufacturing, the volume of material, manual labor, transportation, and the logistics of one-off projects renders the act of building a costly and high-risk business. Ideally, one would hope local craftspeople could be engaged productively. Although debatable, even if Kundoo's labor-based activities are a workable solution for India, in many regions of the world, such custom work is either time or cost-prohibitive, and in some cases, both.

The built environment seems to be at an impasse, today's challenges are unaddressed by maintaining the status quo. Simply designing an amazing building, developing a new design style, or incorporating traditional labor practices will not address seemingly systemic, global challenges. Architects must engage more directly with more efficient production paradigms and somehow manage not to lose ethical and aesthetic design approaches or local artisanal material practices. This dilemma is nothing new. In the early twentieth century, the Deutscher Werkbund, a loose affiliation of architects, artists, and industry groups, perceived tension between mass-produced, inexpensive items lacking critical design consideration and design approaches detached from industrial production that were advocated by the architectural academy that served only the bourgeoisie. The Werkbund evolved from the preceding Jegendstil movement, which privileged ornament and style ahead of production considerations (Posener 1980b, 24) and the UK's Arts and Crafts movement that stood in opposition to industrial means of production. Distinguishing itself from these two movements, the Werkbund sought to address questions about industrial production and the standardization of products (Posener 1980a, 7). Recognizing that society had shifted to mass-producing products, the Werkbund's members questioned how a single design for items such as tableware could benefit a large, diverse group of people by providing well-designed utility (Posener 1980a, 7).

Making architectural designs scalable for mass production, however, introduced additional challenges. The Bauhaus School that emerged from the Werkbund illustrated two different takes on this from its first two directors. Walter Gropius developed custom architectural designs leveraging industrial methods of prefabrication, while Hannes Meyer, where possible, utilized readily available products within architectural assemblages (Gropius & Shand 1965). Meyer's approach was more economical, yet Gropius' more conducive to unique design responses. Although both were scalable solutions to design in an industrial age, with today's robotic fabrication capabilities, it is easy to imagine infinite variations of both Gropius' overall designs and Meyer's industry-produced parts being advantageous. Beyond this, even much of the

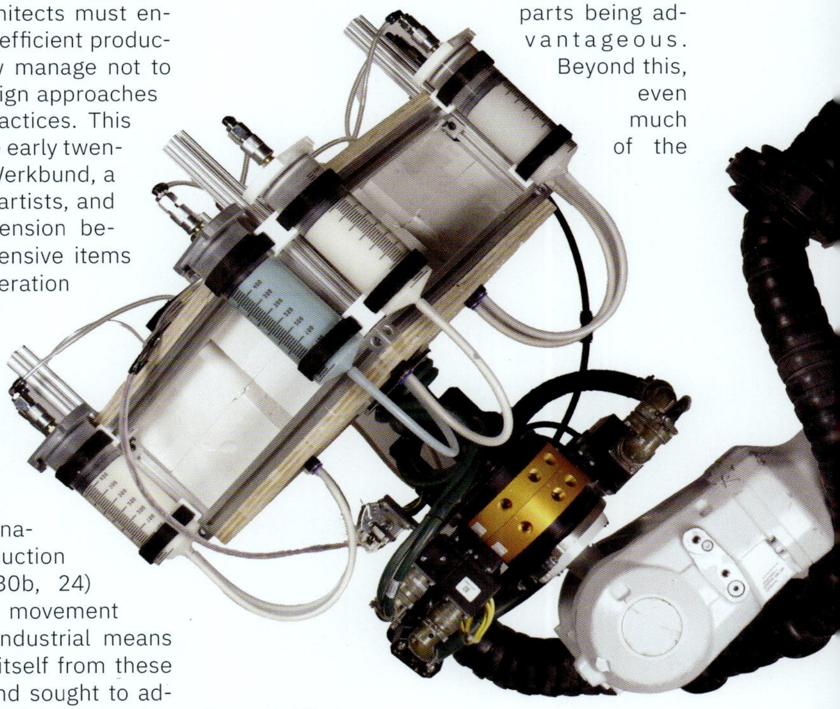

Jegendstil's interest in ornament and figuration can now be produced efficiently using industrial robotic production, rendering both Gropius and Meyer's minimalist approaches as a technological limitation of the past that need not be adhered to with present manufacturing capabilities.

*Robotic "smooshing" manipulation of wet, recently additively manufactured clay (see Robotic Prometheus project, spring semester).*

In this context, a human-centered approach,- such as that of Kundoo or others which supports local artisanal practices while remaining aspirational, echoes the desire of the Arts and Crafts movement, to move away from industrial forms of production. In doing so, such approaches do not seem capable of offering a rapid, scalable solution to high-density, high-quality housing that can be produced efficiently to meet India's urban growth needs. Equally, Gropius and Meyer's approaches arguably lack the cultural or human-centered importance of Kundoo's work. Architectural designs that leverage robotic fabrication offer one means to bridge the divide between these seemingly opposed approaches to design and production. But is robotic production a global possibility or an exclusive opportunity for wealthy economies?

While McKinsey's research suggests that high building demand and high-labor cost countries stand to benefit the most from incorporating robotics and automation, even smaller or emerging economies also stand to benefit to a degree (McKinsey Global Institute 2017, 11–12). Despite India's labor costs being comparatively low, research indicates that incorporating robots into construction for some specific tasks would create a 60% reduction in time, cost, and rework compared to current manual labor practices (Prasath Kumar et al. 2016). Countries currently witnessing a rising demand in the volume, time, cost, quality, or sustainability of construction are even more likely to adopt robotics and automation technologies. Of these, the US, Australia, Belgium, Singapore, China, and the UK in particular, also have limited access to immigrant labor to offset labor costs. They also have access to new technologies and contain a healthy mix of large established companies and disruptive new startups, suggesting these countries are poised for rapid growth in robotic construction capabilities (McKinsey Global Institute 2017, 11–12).

Historically, an expansion in the development and adoption of industrial production capabilities has been instrumental to the marginalization and disappearance of artisanal work and local trade practices. Although recent developments cannot rewind time, they do enable the reintroduction of some lost knowledge in different ways. We are in the early stages of the Fourth Industrial Revolution (IR4), whose key technologies of artificial intelligence (AI), cyber-physical systems, and autonomous manufacturing are set to cause significant disruption and growth. While IR4 risks increasing levels of consumption, its use of autonomous robotics coupled with computational design methods offers a

*Final ceramic assembly prototype (Robotic Prometheus project, spring semester). Students develop designs based on their research project and plan an intensive manufacturing period to produce such large-format prototypes.*

means to bring creativity and artisanal thinking back into industry at scales of production that might meet societal needs, cost-effectively.

## Industrial Revolutions and Design

IR4 allows for greater degrees of user customization, potentially offering more sensitive, thoughtful, and custom-tailored design solutions that might better address the needs of individuals and communities. Coupled with a more eco-ethical mindset, greater specificity in design might also address other species, microclimates, or other factors that buildings directly or indirectly impact. IR4 technologies can also potentially be made more accessible, incorporating more democratic oversight and use and, in some cases, operating as a distributed local infrastructure. To grasp how this is a radical departure from the present, it is important to contextualize these possibilities relative to architecture's historical developments in previous Industrial Revolutions, which each impacted manufacturing and the possibilities for building design in fundamental ways.

Most buildings built today utilize mass production methods first developed in the nineteenth and early twentieth century's Second Industrial Revolution (IR2). In contrast to artisanal work, which required extensive human labor to create unique parts, mass production drastically reduced costs and increased production output by producing vast quantities of identical products using an assembly line or similar process of production. Its economic advantages only operate if production volumes are high enough to offset task-specific machinery and factory overhead costs, rendering variation in production infeasible. Such limitations create a one-size-fits-all approach that encourages generalized design solutions. In buildings, mass production resulted in standardized, prefabricated parts, that ensured a minimum level of quality but also enforced generalized solutions that could only be varied in configuration rather than kind, without sensitivity to diverse building typologies, communities, or climates. Architectural design and discourse aligned with such mass-production approaches in both professional and academic-initiated work. The Deutscher Werkbund and Staatliches Bauhaus (Bauhaus School) mentioned earlier were instrumental in connecting designers to industrial manufacturers and in attempting to unify traditional crafts and creative pursuits with mass production. While they demonstrated significant creativity coupled with the efficiency of mass production, the international style that emerged from this discourse led to a vast amount of building design

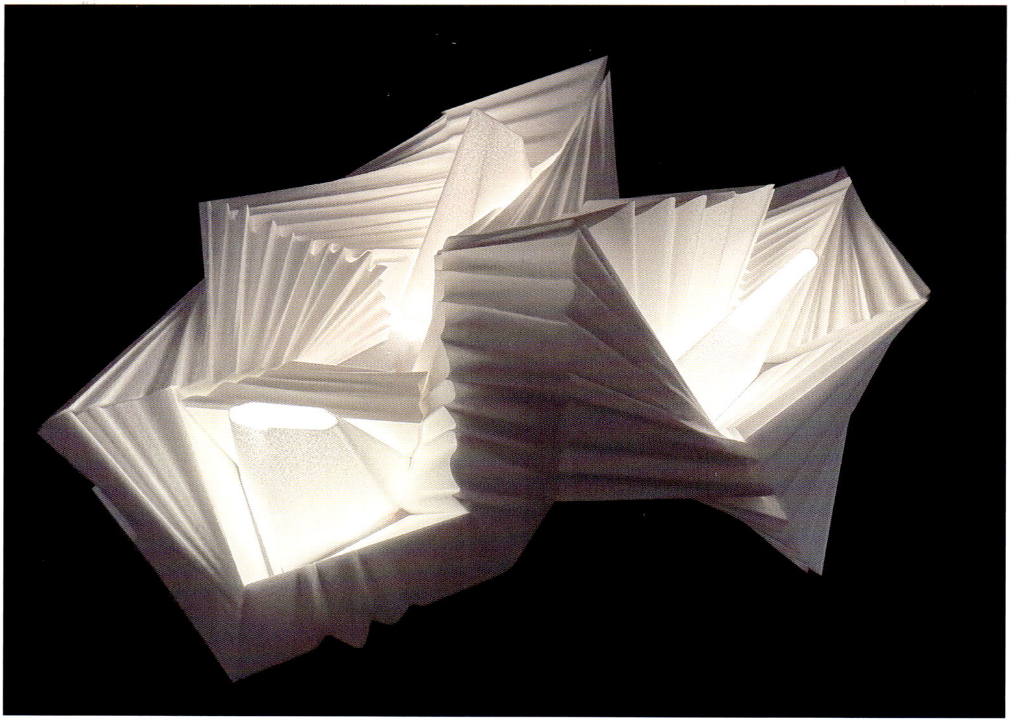

*Above) Fall semester Section 1 partial deep-relief ceiling assemblage prototype and right) rendered view of more extensive ceiling scape (students: Xiangguo Cui, Siyu Dong, Davis Dunaway, and Jingjing Yan).*

replication across cultures and environments, devoid of the artisanal input embodied in earlier periods. While this aided the rapid rebuilding of housing in post-war Europe, there were other aspects of architecture that were subverted by this mode of production. Chicago renders this quite transparently, with designs by Mies Van Der Rohe or his protégés standing near buildings designed by Louis Sullivan. Despite the openness and spatial flexibility embodied in the minimalist work of Mies and co., their projects' near-identical designs also lack the polyscalar ornament of Sullivan's projects, embodying a loss of knowledge and craft that Sullivan leveraged to produce specificity in each project.

In the late twentieth century, the Third Industrial Revolution's (IR3) development of information technology systems, such as computers and the Internet, supported the development of mass customization, enabling the production of geometrically varied parts. Manufacturing incorporated digital file-to-factory workflows and was augmented by CNC machines, laser cutters, articulated robots, and other forms of digitally informed fabrication. Although the automated production of geometrically variable parts was now possible, in practice it was only economically viable for large-scale buildings where the number of parts brought economies of scale or in high-budget projects where cost was less of a constraint, such as in Frank Gehry's Guggenheim Museum in Bilbao (Brayer 2015). Curiously, outside of such seminal works, a different form of design duplication became evident in lesser-known projects. Ironically, information technology that enabled architects to shun the replication of mass-produced designs also led to the mass-sharing and opportunistic adoption of open-source software and algorithms that produced an altogether different form of design homogeneity. As computational knowledge and capabilities have continued to grow in diverse directions there is now arguably less homogeneity in computationally driven design endeavors.

In contrast to the Second and Third Industrial Revolutions, in the nascent Fourth Industrial Revolution, autonomous manufacturing technologies enable bespoke, unique designs at industrial scales of production. In many cases, newfound design freedoms using IR4 methods come at no additional cost. For instance, two additively manufactured buildings of comparable material volume and production time would be comparable in cost regardless of their geometrical design differences. Each could be practically unique, as the manufacturing method, by its nature, treats each project as a one-off production process. Given the early stages in which IR4 is currently, limited exploration has been undertaken into the architectural potentials of such economically competitive design diversity, or its ability to embrace socio-cultural or environmental differences ethically or in an environmentally sustainable way.

It is an exciting time to be venturing into the design and construction industries if one is versed in state-of-the-art, IR4-associated technologies. Construction already accounts for 13% of the world's GDP and is trending upwards. While architects grumble about low salaries, improvements in productivity could create an additional $1.6 trillion a year in economic growth (McKinsey Global Institute 2017, iii). If architects expanded their agency to meaningfully engage in this shift in productivity, they might participate in the benefits. In the short term, China, the US, India, Indonesia, the UK, Mexico, Canada, and Nigeria together represent almost 70% of estimated global construction growth up to 2030, although it really is a global phenomenon (Betts et al. 2015, 15). Culturally diverse and intercontinental in nature, this growth is not limited to a specific national level of economic development. Instead, it is fueled by population growth, rural-to-urban migration, or historical underinvestment, such as in the case of the UK, which is currently short of 4.75 million homes (Betts et al. 2015, 15). With such a diverse set of countries and several reasons driving demand, it is unlikely this global need will disappear anytime soon.

Although the outlook for growth in the construction market is high, the industry also suffers from extremely low corporate investment in research and development. At first appearances, this might appear to be a bad omen. However, this gap leaves substantial space for disruptive ventures that leverage IR4 technological developments. In just over a decade, construction tech has also exponentially gained venture capital (VC) investment at a far greater rate of growth than other VC funded areas –15 times greater than the overall VC industry (Bartlett et al. 2020). VCs know there is a huge upside to be gained by updating our approaches to architecture and construction. Many new building-tech companies are already taking advantage of this, as evidenced by the broad number of concrete

additive manufacturing construction companies that have recently emerged such as US' Icon, China's Winsun Decoration and Engineering Co, Russia's Apis Khor, or France's XTree3D, amongst others. Metal additive manufacturing (Wire Arc Additive Manufacturing or WAAM) has also made its debut in the work of MX3D in the Netherlands (MX3D 2020), alongside rapid robotic bricklaying in Australia by Fastbrix's Hadrian X robot (FBR Ltd 2018). On-site construction activities are also now being supported by several task-specific robots from companies including the US' Boston Dynamics and Dusty Robotics, alongside China's Guangdong Bozhilin Robot Co Ltd (iNEWS 2022) and road construction Xuzhou Construction Machinery Group (Rogers 2021), amongst others.

Several robotic fabrication companies have also entered the market, including Blueprint Robotics, Intelligent City, and AutoVol, that prefabricate timber buildings, through to individual trades such as Quarrestone for stone-cutting or Mesh Mould for non-standard steel reinforcement fabrication. Several creative or technical start-ups have also been launched by young architects including companies such as AUAR, Nagami, AiBuild, and others that offer design and manufacture of mass timber buildings, thermoplastic façades and interiors, or sell robot manufacturing and design software stacks, or in the case of Automata, novel industrial robot systems. Many architects have also been recruited into established tech companies such as Boston Dynamics, Wework, the late Google Sidewalks Lab, and others. A small number of architectural practices are also actively realizing robotically fabricated designs, such as Phillip Yuan's Archi-Union Architects in China (Wu et al. 2022), Snooks+Harper in Australia (Snooks+Harper 2023) or Studio Rap in the Netherlands (Barandy 2023), amongst others. In this context,

*substantial creative and technical agency exists for those who venture into a rapidly changing construction industry that is presently open to novel ideas, workflows, and companies.*

## Architectural Robotics Education

Over the last two decades, an architectural robotics field of research has emerged and gained traction in architecture schools internationally. Since Gramazio and Kohler's introduction of industrial articulated robot arms in teaching and research at ETHZ in 2005 (Gramazio et al. 2014), and Achim Menges and Jan Knippers robotically fabricated pavilions at University of Stuttgart from 2011 to present (Menges & Knippers 2011), a large number of architectural schools have integrated robotics into research and teaching. Early adopters include IAAC and University of Innsbruck in Europe, University College London, and the Architectural Association in the UK; RMIT, UTS, UC in Australia; Tonjii and Tsinghua Universities in China; University of Pennsylvania, University of Michigan, Harvard GSD, Princeton University, UCLA, SCI-ARC, Pratt Institute, University of Texas, and others in the US. The actual number of institutions with these general-use robots is substantially greater than those mentioned. The diversity and breadth of this community's research is captured in international conferences such as RobArch, Fabricate, ACADIA, SIIGRAPH, IROS, ICRA and others.

*Despite a broad and growing interest in architectural robotics, there are few educational offerings in the US that are focused on architectural design approaches that leverage robotic fabrication or that address the challenges and opportunities the Fourth Industrial Revolution affords in building design and production.*

To do so, requires an institution to make a substantial capital investment in robotics-based facilities, staffing, and new academic programs in order to have sufficient scope and capabilities to explore challenging intellectual and practical questions around robotic fabrication and design in parallel. It is, however, towards these questions that the University of Pennsylvania Weitzman School of Design's Masters of Science in Design: Robotics and Autonomous Systems program (MSD-RAS) is focused. The MSD-RAS program seeks to develop novel architectural agencies and outcomes through

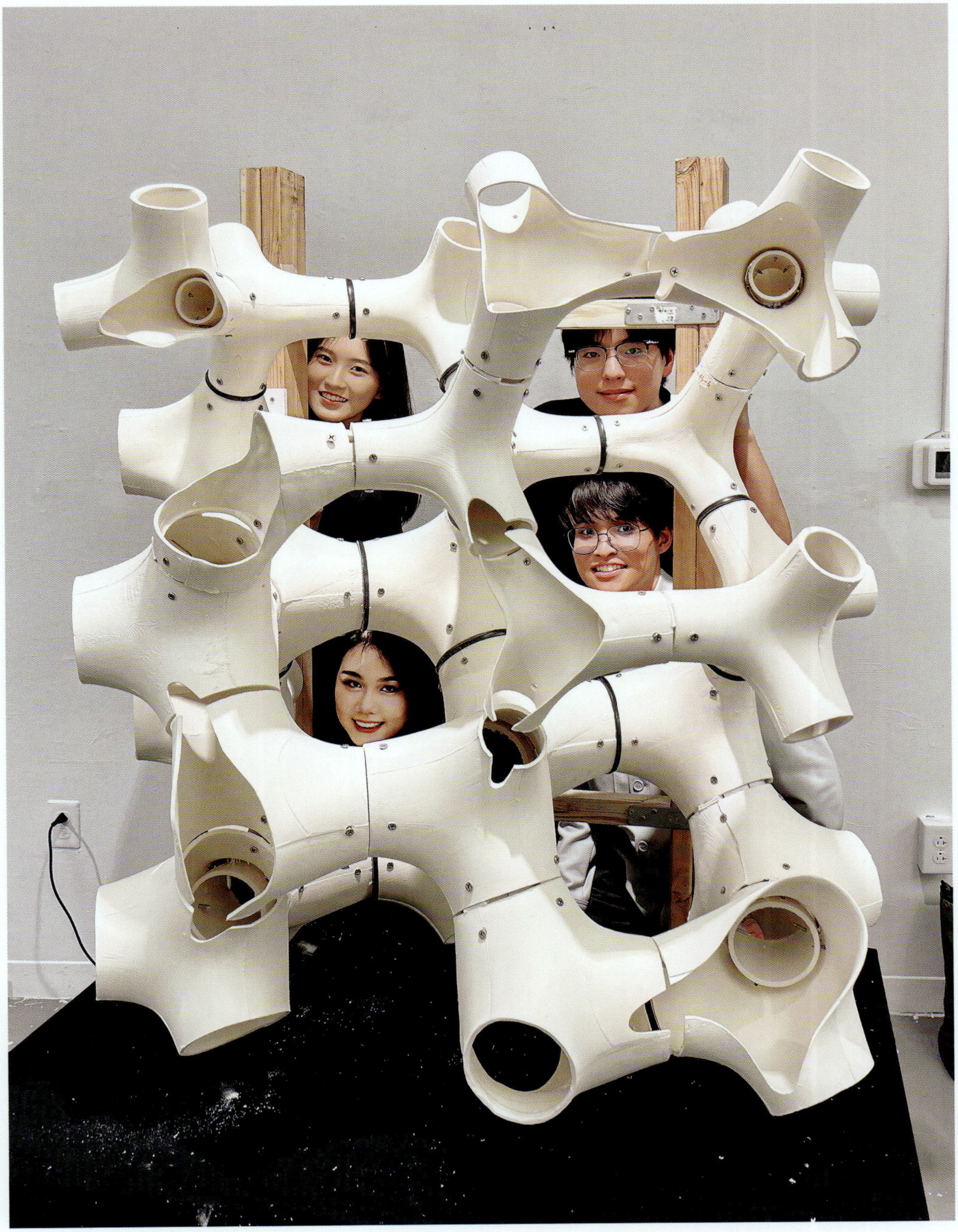

*A team of students during the last semester of the MSD-RAS program. Their final ceramic assembly prototype was temporarily assembled prior to parts being individually glazed (Dynamic Slip-Casting project, spring semester)*

research and experimentation into robotic fabrication and computational approaches to design. While robotic fabrication is usually concerned with automated means of manufacturing, the MSD-RAS is distinguished by its emphasis on Fourth Industrial Revolution capabilities in autonomous manufacturing, due to their flexible design and manufacturing possibilities.

While automated systems operate unaware through predetermined routines, autonomous robot systems are programmed to be adaptive to dynamic events. They can incorporate live sensor data feedback to engage with people, objects, or the environment. Autonomous systems can therefore extend design agency into spatiotemporal events and embody more direct relationships between end-users, designers, and fabricators. Many of the computational workflows explored in the program are also semi-autonomous in themselves, several falling under the umbrella term "Artificial Intelligence" (AI), such as generative design, machine learning, reinforcement learning, and multi-agent systems, amongst others.

In the MSD-RAS program, design agency operates through robotic fabrication and AI approaches to computation to engage sensitively and distinctively with a broad range of physical media to produce bespoke outcomes that can also be theoretically scaled to high volumes of production without compromising their ability to produce nuanced one-offs. This focus on integrated computational approaches to design and manufacturing aims to unite conception and realization activities into a cohesive, creative endeavor that addresses logistical, material, and aesthetic considerations simultaneously. The MSD-RAS is not solely delving into robotic techniques in insolation. The program also engages in a critical examination and reimagining of industrial and artisanal methods for materials and manufacturing in order to enhance their abilities to support design outcomes that are more varied, sensitive, and specific.

As an Ivy League educational offering, the Weitzman School of Design's MSD-RAS program does not merely offer technical training in robotic and computational workflows. The program has embarked on a critical yet creative approach to Fourth Industrial Revolution design research that engages with architectural discourse, posthuman and environmental ethics and aesthetics, the socio-politics of design and making, together with the history and theory of computation and robotics, as well as providing philosophical perspectives such as Object Oriented Ontology, Vital Materialism, Material Agency, and others. Despite the perception that robotic fabrication and computational

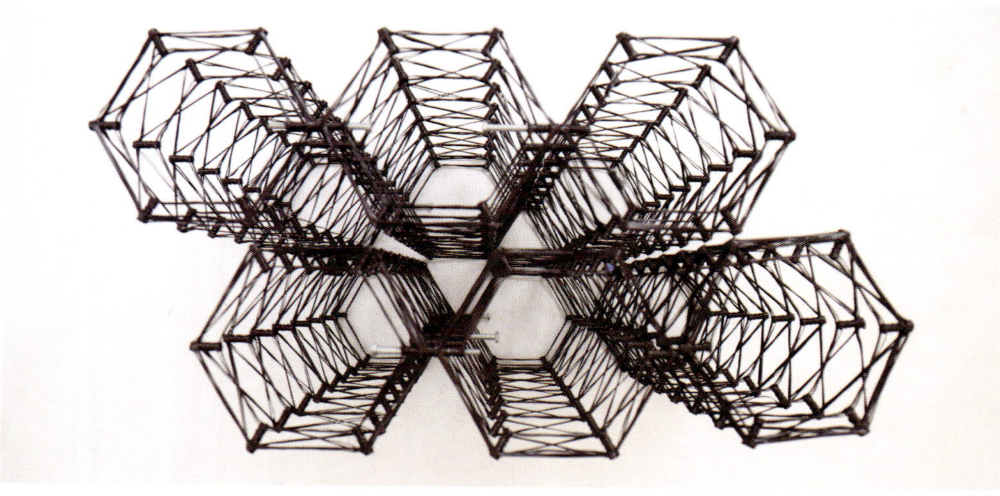

*A lightweight Fall Semester Section II assembled prototype comprised of robotically filament-wound, pre-impregnated carbon fiber strands (See Fall Section 2, students: Grey Wartinger, and Jiansong Yuan).*

design are exclusively technical in focus, these activities are arguably where some of the most novel creative design activities are taking place. This is happening particularly in areas of research that are leveraging deep learning and other forms of AI together with adaptively programmable means of robotic fabrication and construction. The research and teaching covered in the MSD-RAS offers new approaches to the ideation and realization of design.

*The program thus aims to be industry-relevant, spurring diverse career opportunities whilst remaining creative and open enough to pursue exploratory, pioneering research and speculative design scenarios. To achieve these ambitions uniquely, the RAS curricula is built around integrated approaches to the design and production of robotically fabricated prototypes.*

Full-scale, multi-part assembled prototypes are developed in all three of the program's design studios, each developed through applied learning from RAS courses that run in parallel to support these endeavors. Through this hands-on approach to knowledge and skill building, an ambitious curricula offers an unprecedented opportunity to question and develop possible futures for architectural design and production. A lot of thought has gone into the program's unique pedagogical structure, so much so, that we felt compelled to share more about it in the essay that follows this introduction.

The MSD-RAS students' work has also made a valuable contribution to academic research in the broader design community. Several peer-reviewed papers were accepted for publication during this time in world-leading conferences, demonstrating that the work of the program is novel and making an impact in academia. It is remarkable to witness graduates publishing and presenting 16-week projects that stand up competitively alongside research from career academics and final-year PhD students. MSD-RAS students are provided with a high-level of support to get to the point where they become active members of the broader architectural research community. This support comes from

*Computational design strategizes material placement (See Fall Section 2, students: Xiangguo Cui, Siyu Dong, Davis Dunaway, and Jingjing Yan).*

deeply commited faculty who are all leaders in their respective fields in both industry and academia, including:

***Andrew Saunders,*** Associate Chair of Architecture and Director of Penn's M.Arch program, has years of experience in robotically fabricated design projects and pavilion projects, most recently the Deep Relief Feature Wall at Middletown Free Library featured towards the end of this book (Saunders et al. 2022). Saunders leads one of the fall Material Agencies' design studios and is the author of several books including *Baroque Topologies* and *57 Pavilions* (Saunders 2018; Saunders & Dubbeldam 2018).

***Evangelos Kotsioris***, instructor for the fall semester RAS Theory course, holds a PhD in History and Theory from Princeton University and is an architectural curator at the world-renowned Museum of Modern Art in New York. Dr.Kotsioris' knowledge of cybernetics and the history of architecture and computation has provided a strong foundation for RAS students embarking on design research in the program.

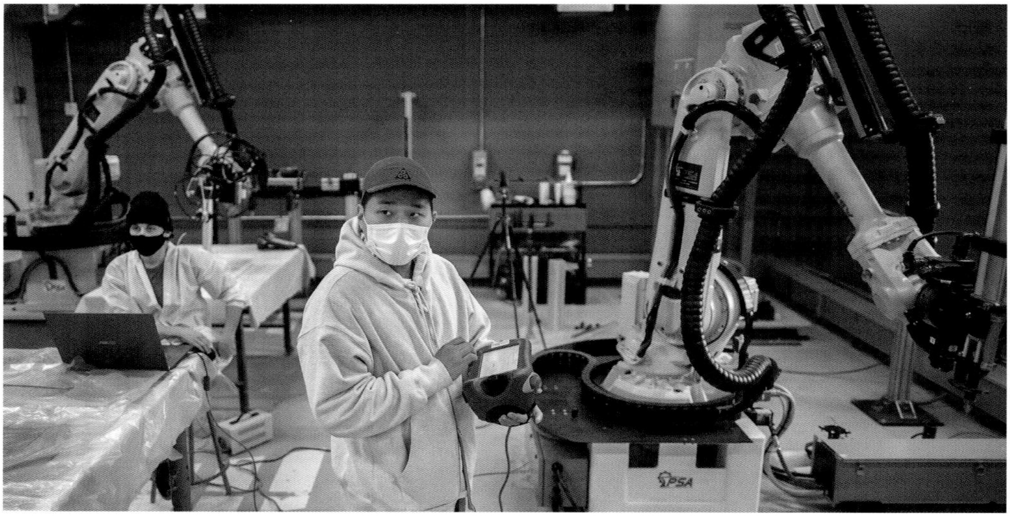

*Students spend most of their time on project-based work in the state-of-the-art ARI Robotics Lab facility located in Weitzman School of Design, right at the heart of Penn's campus.*

***Nathan King*** is the industrialized construction research lead at Autodesk, where he develops applied research within architecture, engineering, and construction and contributes to the development of strategic initiatives related to robotics, automation, and the future of construction. Dr.King brings state-of-the-art industry knowledge to spring's Experimental Tooling course, including ceramic production approaches published in his co-authored book *Ceramic Material Systems: in Architecture and Interior Design* (Bechthold et al. 2015). He also holds a doctorate from Harvard GSD's DDes program.

***Billie Faircloth*** is an adjunct professor at Penn and partner in the architectural practice KieranTimberlake. She leads a transdisciplinary group of professionals leveraging research, design, and problem-solving processes from fields as diverse as environmental management, chemical physics, materials science, and architecture. Professor Faircloth fosters collaborations between disciplines, trades, academies, and industries in order to define problem-solving boundaries for the built environment. She is also author of the book *Plastics Now: On Architecture's Relationship to a Continuously Emerging Material* (Faircloth 2015). Professor Faircloth brings substantial academic and practice-based research experience to the MSD-RAS spring semester Scientific Research and Writing course.

***Jeffrey Anderson*** teaches fall's Introduction to Cyberphysical Systems and spring's Advanced RAS Programming courses. Anderson is an educator, architectural designer, and AR/VR software developer. As a lecturer at Penn, he is also the lead software developer in the Design Lab at Mancini Duffy, where he conducts design research and develops architectural visualization tools. His research focuses on using technology to create new relationships between users, architecture, and context through interaction, sensing and feedback, and mixed reality. He is also co-author together with Alejandro Zaera Polo for the recently published book *The Ecologies of the Building Envelope: A Material History and Theory of Architectural Surfaces*, that analyzes how social, political, technological, and economic forces have become embedded within architecture over the last century (Zaera-Polo et al. 2021).

***Ezio Blasetti*** is a senior lecturer at Penn and teaches the second MSD-RAS fall Material Agencies' design studios and the fall Algorithmic Design and Robotic Fabrication Course. He is a founding partner of the architectural practice Maeta Design and Apomechanes, an annual intensive design workshop exploring algorithmic processes and fabrication. Blasetti has extensive experience in computational design and robotic fabrication, bringing a lot of hands-on knowledge to his MSD-RAS courses.

***Patrick Danahy*** teaches the fall RAS Algorithmic Design and summer Introductory courses that focus on machine learning approaches to computational design and robotic fabrication, such as Deep Reinforcement Learning. Danahy is a lecturer at Penn and an assistant research professor at the University of Texas. As both a designer and programmer, Danahy is a leader in AI approaches to generative design.

***Jose Luis García del Castillo y López*** also co-taught spring's Advanced RAS Programming course in its first year. García del Castillo y López is a lecturer at Harvard GSD and is the developer behind the Robot ExMachina software framework that enables real-time programming on several different industrial robot platforms. He is also the founder of the YouTube channel ParametricCamp, a computational design learning resource. García del Castillo y López's research offers a pathway towards more participatory, engaging human-robot collaboration and real-time design and fabrication.

Some exciting new faculty also commenced this year whose courses could not be included in this book. Alicia Nahmad Vazquez is running one of the fall Material Agencies design studios. Dr.Nahmad Vazquez is an innovator in robotic timber fabrication as an academic and entrepreneur, an associate professor at the University of Calgary, and founder of The Circular Factory and MIToolbox. Dr.Nahmad Vazquez is both a designer and robotic fabricator, with a wealth of intellectual and practical knowledge that drives her work and teaching. Emek Erdolu is now running the fall RAS Theory course, bringing a wealth of knowledge from his practice experience in the offices of AECOM, HMD, Ecoland, and Nomad Studio, and as a researcher in the ETH Future Cities Laboratory (FCL). Emek's PhD in Computational Design at Carnegie Mellon University (CMU) School of Architecture investigated how building tasks are analyzed and decomposed to inform on-site robotic systems and how such systems are organizing new workflows and human-machine interactions.

## About this Book

This book is the first to cover work from the MSD-RAS program. It is organized primarily around student projects that are presented chronologically in the sequence in which they occurred in the program's academic year, which included two fall projects and one spring semester project. Dispersed amongst this work is a series of discrete but interconnected essays written by some of the faculty on topics explored in the program that can be read in any order.

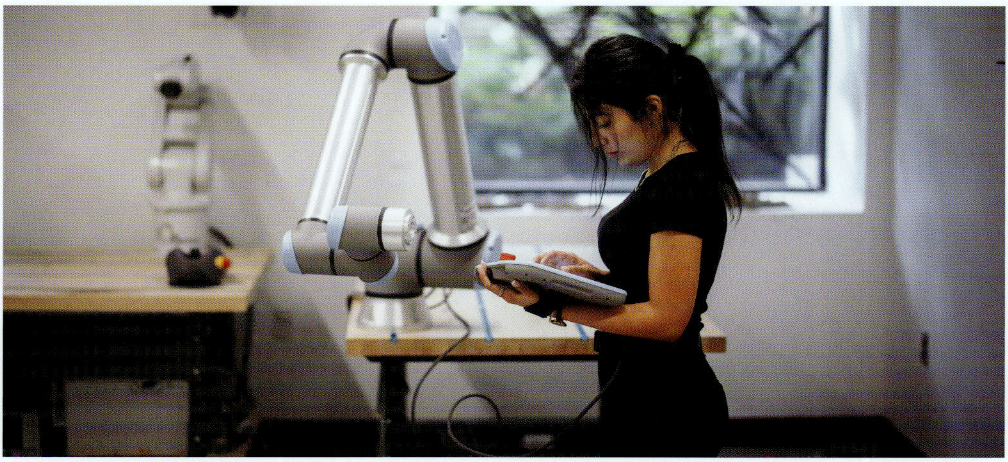

*Students learn to program and fabricate with a variety of different industrial robots.*

The collaborative spirit embodied in this book is not only apparent in these student team projects but also in the workings of the MSD-RAS program itself, which operates through an ambitious pedagogical structure. The first essay discusses the program's unique integrated approach to curricula and project work and how this supports competitive cutting-edge research outcomes and exciting career possibilities. Following this, ***Billie Faircloth*** discusses what it means to pursue "design research" that is not solely design work or solely research. She also examines the role of scientific writing in the MSD-RAS curriculum, and the program's connection to the broader design research community. Leaning into the spring semester's shared research theme on architectural ceramic assemblies, ***Nathan King*** explores the messy, inprecise aspects of clay in constrast to the precision of robotics, situating design research in relation to industry mass-production methods and artisinal processes. ***Jeffrey Anderson*** then writes on human-robot interaction and an integrated approach to design and production. While it might come as a surprise that a program exploring autonomous systems might devote attention to human-in-the-loop collaborative frameworks, there are distinctive capabilities in human and robot agencies that warrant the development of collaborative approaches to design and production.

Tying these themes together, the author then discusses the program's integrated approach to design and how it extends design agency into other media such as materials, computational processes, robot tools, robot platforms, etc. In the MSD-RAS, some form of ownership is assumed of material and production concerns, in order to produce highly specific design possibilities through custom-developed approaches that reduce waste and material use while also being theoretically scalable to industrial production volumes.

Towards the end of the book, graduates also discuss their own experiences as both students and as graduates forging exciting careers. Data from student surveys also reveals the degree to which knowledge gained in the program is applied in these careers and provides insights into graduates' salaries. The book concludes with a series of exciting faculty research projects MSD-RAS students significantly contributed to during, or directly after, their studies in part-time or full-time employment at the Weitzman School of Design. These projects ran adjacent to the MSD-RAS curriculum and formed part of the research environment students participated in during their studies. A series of course descriptions are also provided as an appendix. Together, this diverse content aims to offer a glimpse into the motivations behind the program, the way in which it operates, and its novel research outcomes.

*Robotic Fabrication and Architectural Design* is the first book on Penn's MSD-RAS program. The content is focused on a series of robotic fabrication and material production approaches that were explored creatively and discursively during the first three years of the program, the first two of which experienced interruptions from the COVID-19 global pandemic. It is incredible to see how much was achieved in this new program during such a disruptive period. This is ultimately a snapshot of an initial set of projects and research themes explored during this time. MSD-RAS' research themes and courses are constantly evolving. At the time this first edition is being published, the program is already engaged in entirely different material, fabrication, and research themes across its three design studios and other courses. These are being explored within a new multi-year research agenda that will likely be the subject of a follow-on book. This book provides a glimpse into the program's hive mind, its collective thinking around a series of architectural and manufacturing concerns that seek to extend architectural design inquiries beyond blind technological adoption, to think critically and creatively into new design possibilities that leverage state-of-the-art technologies – particularly in the fields of robotic fabrication and computational design.

## References

Barandy, K. "Studio RAP Completes 3d-Printed 'Ceramic House' in Amsterdam." *Designbloom*. November 30, 2023. https://www.designboom.com/architecture/studio-rap-ceramic-house-amsterdam-with-3d-printed-facade-11-30-2023/.

Bartlett, K., J.L. Blanco, B. Fitzgerald, J. Johnson, A.L.

Mullin, and M.J. Ribeirinho. "Rise of the Platform Era: The Next Chapter in Construction Technology." Mckinsey & Company. October 30, 2020. https://www.mckinsey.com/industries/private-capital/our-insights/rise-of-the-platform-era-the-next-chapter-in-construction-technology.

Bechthold, M., A. Kane and N. King. Ceramic Material Systems: In *Architecture and Interior Design*. Basel: Birkhäuser, 2015.

Betts, M., G. Robinson, C. Burton, J. Leonard, A. Sharda, and T. Whittington. "Global Construction 2030: A Global Forecast for the Construction Industry to 2030." *Global Construction 2030*. 2015 www.globalconstruction2030.com

Bose, S. "Anupama Kundoo and the Architecture of Happiness." *Wallpaper*. August 17, 2022.https://www.wallpaper.com/architecture/anupama-kundoo-profile

Brayer, M.A. *Frank Gehry: The Interlacing of the Material and the Digital*. Edited by A. Lemonier and F. Migayrou. Munich: Prestel, 2015.

Desai, M. *Women Architects and Modernism in India: Narratives and Contemporary Practices*. New York: Taylor & Francis, 2016.

Faircloth, B. (2015). *Plastics Now: On Architecture's Relationship to a Continuously Emerging Material.* London: Routledge, 2015.

FBR Ltd. "Hadrian X: Outdoor Construction and Bricklaying Robot from FBR." 2018. https://www.fbr.com.au/view/hadrian-x

Gramazio, F., M. Kohler, and J. Willmann. "Authoring Robotic Processes." *Architectural Design* 84, no. 3 (2014): 14-21. DOI. https://doi.org/https://doi.org/10.1002/ad.1751

Harrouk, C. "Awareness Makes Us Human: In Conversation with Boonserm Premthada, Founder of Bangkok Project Studio." ArchDaily. February 12, 2023. https://www.archdaily.com/987201/awareness-makes-us-human-in-conversation-with-boonserm-premthada-founder-of-bangkok-project-studio

Heringer, A. (n.d.). "Vision." Anna Heringer Architecture. Accessed on June 4, 2024. https://www.anna-heringer.com/vision/

International Energy Agency (IEA). "2019 Global Status Report for Buildings and Construction." date, 2019.

International Labour Organization. "Construction: A Hazardous Work." March 23, 2015. https://www.ilo.org/global/topics/safety-and-health-at-work/areasofwork/hazardous-work/WCMS_356576/lang--en/index.htm

Kundoo, A., E. Heathcore, M.M. Kallehauge, and L.R. Jørgensen. *Anupama Kundoo: The Architect's Studio*. Baden: Lars Müller Publishers, 2020.

McKinsey Global Institute. "Reinventing Construction: A Route to Higher Productivity, Executive Summary." date, 2017. https://www.mckinsey.com/capabilities/operations/our-insights/reinventing-construction-through-a-productivity-revolution

Menges, A. and J. Knippers, "ICD/ITKE Research Pavilion 2010." University of Sttugart. 2011. https://www.icd.uni-stuttgart.de/projects/icditke-research-pavilion-2010/.

MX3D. "MX3D Bridge." MX3D. 2020. https://mx3d.com/industries/mx3d-bridge/.

Osmani, M. . "Construction Waste." In *Waste: A Handbook for Management,* edited by Trevor M. Letcher and Daniel A. Vallero, 207-218. Cambridge: Academic Press, 2011. . https://doi.org/10.1016/B978-0-12-381475-3.10015-4

Plumb, T. "IMF: AI Will Impact 60% of Jobs and Worsen Inequality." VentureBeat. January 18, 2024. https://venturebeat.com/ai/imf-ai-will-impact-60-of-jobs-and-worsen-inequality/

Posener, J. "Between Art and Industry: The Deutscher Werkbund." In *The Werkbund: Studies in the History and Ideology of the Deutscher Werkbund, 1907-1933*, edited by L. Burckhardt, 7-15. Woodbury: Barron's, 1980a.

_______. "Werkbund and Jugendstil." In *The Werkbund: Studies in the History and Ideology of the Deutscher Werkbund, 1907-1933*, edited by L. Burckhardt, 16-24. Woodbury: Barron's, 1980b.

Prasath Kumar, V. R., M. Balasubramanian, and S. Raj Jagadish. "Robotics in Construction Industry." *Indian Journal of Science and Technology* 9, no. 23 (2016). https://doi.org/10.17485/ijst/2016/v9i23/95974

Rogers, D. "'Landmark Event' Claimed as Robots Pave Busy Highway in China." *Global Construction Review*. October 11, 2021. https://www.globalconstructionreview.com/landmark-event-claimed-as-robots-pave-busy-highway-in-china/

Saunders, A. *Baroque Topologie*s. Rome: Palombi Editori, 2018.

Saunders, A. and W. Dubbeldam, *57 Pavilions*. Novato: ORO Editions, 2018.

Saunders, A., R. Studebaker, and C. Eileen Moriarty. "Deep Relief." In *Hybrids and Haecceities: Proceedings of the 42nd Annual Conference of the Association for Computer Aided Design in Architecture*, edited by M. Akbarzadeh, D. Aviv, H. Jamelle. and R. Stuart Smith, 652-661. La Vergne: IngramSpark, 2023.

UN DESA. "World's Population Increasingly Urban with More than Half Living in Urban Areas." United Nations Department of Economic and Social Affairs. July 10, 2014. https://www.un.org/en/development/desa/news/population/world-urbanization-prospects-2014.html

UNHCR. "Refugee Statistics." USA for UN Refugee. May 2023. https://www.unrefugees.org/refugee-facts/statistics/

Wainwright, O. "'It's Already Way Beyond What Humans Can Do': Will AI Wipe Out Architects?" *The Guardian*. August 7, 2023. https://www.theguardian.com/artanddesign/2023/aug/07/ai-architects-revolutionising-corbusier-architecture?CMP=Share_iOSApp_Other

Wu, H., M. Lu, X. Zhou, and P.F. Yuan. "Application of 6-Dof Robot Motion Planning in Fabrication." In *Proceedings of the 2021 DigitalFUTURES*. Singapore: Springer, 2022. https://doi.org/10.1007/978-981-16-5983-6_31

Zaera-Polo, A. and J. Anderson. *The Ecologies of the Building Envelope: A Material History and Theory of Architectural Surfaces*. New York: Actar Publishers, 2021.

# PRIMED FOR SUCCESS:

## *What Makes RAS a Unique Educational Experience?*

Transformative societal and industry changes are afoot with the rise of the Fourth Industrial Revolution (IR4) that will irreversibly impact the architecture and construction industries. As exciting and challenging as this is, architects are not well-equipped for this emerging frontier. Globally, only a small set of academic design programs extensively explore this industrial revolution's societal, technical, creative, and/or discursive implications. An industrial revolution represents a paradigm shift that has the potential to upend established modes of practice and knowledge exchange. To engage effectively with such a transformative subject matter requires an educational model tailored to this new frontier. To address this, in the University of Pennsylvania's Stuart Weitzman School of Design, the post-professional Masters of Science in Design: Robotics and Autonomous Systems degree program (MSD-RAS) is exploring a hyper-collaborative pedagogy that places greater emphasis on the synthesis of knowledge across all courses rather than on an isolated, design-led studio course, integrating advanced technical knowledge within creative design activities that take place within a state-of-the-art new robotics facility: the Weitzman School of Design's ARI Robotics Lab.

The MSD-RAS program explores architectural design through integrated approaches to robotic fabrication and computational design. In the MSD-RAS, learning is centred around the production of a series of experimental fabricated prototypes. This project-oriented approach is used to synthesize diverse curricula through a form of applied learning aiming to empower students with tacit knowledge that can be transformative in their future careers. Both creative and scientific in spirit, it is hoped this pedagogical approach fosters novel and exploratory work while supporting a more diverse group of students and career trajectories for graduates. Run from within the Department of Architecture, the MSD-RAS' novel approach to design education is strategically and practically differentiated from other offerings in design research education. So much so that some aspects of the program's unique structure discussed in this essay were included in a paper recently published at the 2023 ACSA/EAAE Teachers Conference in Reykjavik, Iceland (Stuart-Smith 2023). This essay extends this discourse, also illustrating how the high-calibre faculty and the co-development and management of the RAS curricula and ARI robotics lab facilities propel the program forward, offering a tangibly different form of design-research education that is geared towards training graduates for the careers of tomorrow.

### Educating for Future Careers

Higher Education aims to equip students with the knowledge and skills to empower them to have successful careers that span from the present to decades into the future. This is challenging when socio-economic and technological changes rapidly transform the world graduates operate in and corresponding career prospects. How do we equip students for the unknown? Educational institutions can commence by addressing present and emerging technological and industrial trajectories and by operating both practically and speculatively in parallel.

*Spring semester thesis project jury. Guest critics experienced in the research subject area review the work before a final research paper and thesis book are produced, providing valuable, constructive feedback.*

At this critical juncture with the commencement of IR4 there is growing consensus that rapid developments in robotics, artificial intelligence, and material science are facilitating the emergence of autonomous manufacturing, where economically scalable approaches to variable, distributed manufacturing on-demand, user-customization, and locally sourced or synthesized materials become more possible (Schwab 2017). Coupled with cyber-physical systems, there is the potential for a more direct engagement between the built environment's inhabitants, designers, and manufacturers. While these nascent capabilities offer untold possibilities, they also threaten to increase inequity, consumption, and waste. The growing use of machine learning and other forms of software automation also increasingly challenges established roles within the global workforce including in architectural practice. It seems imperative that architectural education must leverage these recent and emerging developments to increase the architect's agency, and to address pertinent issues such as building affordability, resource and infrastructure equity, environmental impact, and more. Associated societal changes also speak to a cultural need for us to further the aesthetic possibilities of design and its discourse. To achieve this, more technological engagement is needed within architectural education to build bridges between design, material, and fabrication concerns.

In this context, timeless questions reemerge, such as whether architects can or should be developing novel technologies that might offer practical steps towards realizing alternative visions of the future. Or are architects, by vocation, constrained to adopt established products and manufacturing methods with no means to strategically impact the present or future trajectory of the building industry? The discipline's answers to these questions are varied. In this book's introduction, several robotic fabrication and technology start-ups founded by architects were mentioned. Do these expand architectural agency, or did architects switch to different roles, employing different skills when founding those companies? In commercial architectural practice, recent industry trends exhibit fragmentation of the built environment's design, manufacturing, and construction activities into an increasing number of fields of specialization over time. This almost hyper-specialization of roles has eroded some of the architect's historical scope of work, with the architect largely perceived as a generalist. This perception is warranted given that one meets far more PhD graduates in an engineering or specialist consultancy firm than in an architectural practice. It is also much clearer how PhD research contributes to those discipline's core knowledge and service offerings. It is rare for an architectural practice to mirror this level of foundational research in their services. In the MSD-RAS, however, this status quo is questioned in two different ways. Firstly, can we develop new and agile generalist knowledge, and secondly, can we foster new research specializations that are targeted towards specific architectural and construction orientated agencies?

Arguably, any field's scope of work emerges from a domain in which they have authoritative knowledge and skills. As the trend towards specialization implies, these domains seem to be fluid. There is always room to maneuver and stake out new ground. Regardless of where professional accreditation boards demarcate the limits of architectural expertise or the title and job in which this is applied, there are arguably no barriers to how or where one might practice architectural knowledge (e.g. one of the Architectural Association School of Architecture's most cherished alumni is the late Mark Fisher, founder of Stufish, a company that designed concert stage sets for U2, Pink Floyd, and other rock legends).

## Beyond a Vocational Education

Are architectural degree programs primed towards the development of new forms of knowledge and impact, or more towards vocational training? Architecture students graduate with an exceptional range of knowledge and skills. Despite this, a professional architecture degree leads to a limited number of employment opportunities outside of the vocation. This is largely due to a focus on a nationally monitored set of design, construction, and communication key performance indicators that aim to safeguard a professional standard of education (e.g. NAAB (NAAB 2023)). Contrast this with universities' scientific and engineering schools, where more extensive amounts of funded research enables novel developments and start-ups closely aligned to or seeded from academic activities. In comparison, the vocational education of architecture is less well-equipped to proffer disruptive change or help to shape technological advances in industry and practice. Research-led design programs, however, do provide a means to address these concerns.

In the US and many parts of the world, architectural professional degree programs are centered on project-orientated work stemming from eighteenth-century France's L'Ecole des Beaux Arts educational model (Cunningham 1979). As Allen Cunningham describes, the Beaux Arts system placed primacy on the design project, undertaken within an "atelier" run by an experienced "patron" architect (Cunningham 2005). This mode of learning is essentially some form of master-apprentice model that is more focused on imparting historical knowledge than empowering students to develop new, self-discovered knowledge. According to American academic and former President of the Association of Collegiate Schools of Architecture (ACSA), Rashida Ng, the Beaux Arts model most American architectural programs operate under is insular in nature and was designed to cater for a historically elitist, wealthy, white male student. She argues that radical changes should be implemented in architectural pedagogy to support more collaborative open dialogue that is inclusive of all students and recognizes students' diverse backgrounds, be it racial, gender, sexual orientation, social, or economic in nature (Ng 2022). Professor Ng also suggests that the primacy of "design" leaves little room for students to foster other aspects of their architectural education that could lead to more diverse career opportunities (Ng 2022).

A notable alternative approach to architectural education was offered in the Das Staatliche Bauhaus Weimar in Germany. Founded by Walter Gropius in 1919, the Bauhaus compressed both creative and technical workflows within studio workshops, several focused on the production of prototypes suited for potential mass-production. Initially in the Bauhaus' Weimar location, these studios were co-led by a teacher of design and a practical instructor who was well-versed in industrial manufacturing and material production methods. By the time the Bauhaus had moved to Dessault, a generation of graduates were capable of single-handedly teaching this integrated curriculum (Gropius & Shand 1965, p. 95). In both iterations, lessons prioritized practical work and students were reminded of real-world logistics through regular engagement with industry (Gropius & Shand 1965, 77). As this curriculum still focused on the top-down imparting of known knowledge from faculty and industry, it was not necessarily conducive to the fundamental research that takes place in scientific fields within academia, which has the potential to initiate a paradigm shift in industries' practices and outlook. The Bauhaus' hands-on educational model existed for only a brief period of time before being pressured to close in 1933 in the lead-up to World War 2. Despite this, it had a significant impact, with graduates taking up leading roles in industry and academia world-wide. Had it continued, one can assume it would have also sought to move on beyond its interest in mass-production towards IR4 modes of production available today.

While changes must be made to professional degree programs, post-professional Masters and PhD programs do offer opportunities for students to engage in more diverse subject areas and build more career opportunities through the attainment of research-orientated knowledge and skill specialization. Some are also recognized as STEM-certified, acknowledging the role of science, technology, engineering, and mathematics within their curricula (Bair et al. 2018). In the Weitzman School of Design's Department of Architecture, a series of STEM-certified post-professional Masters of Science in Design (MSD) programs offers graduates from professional degrees opportunities to specialize in one of a number of different knowledge streams. The most recent addition to these, MSD-RAS, focuses on the integration of robotic fabrication and computation within design. MSD-RAS aims to fill an education gap where state-of-the-art robotics, fabrication, computation, and design knowledge can add value to the field of architecture by training and empowering graduates to make an impact in industry, practice, and academic research. The degree is also open to students with non-architectural backgrounds who demonstrate equivalent experience, enabling a degree of multidisciplinary collaboration among students. With this ambitious remit, the pedagogical structure of the MSD-RAS program, its high-caliber faculty and

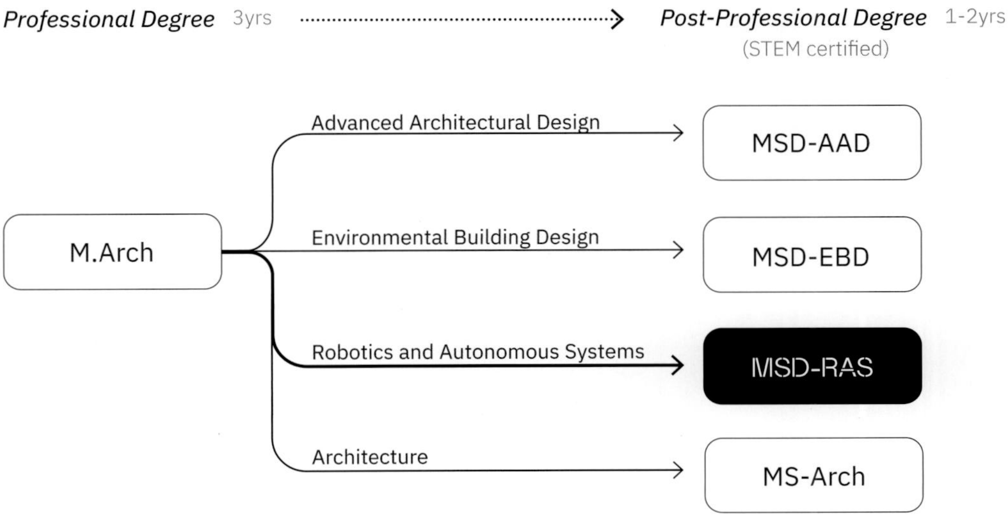


*The MSD-RAS is one of several different specialist STEM post-professional Masters of Science in Design programs offered in Penn's Department of Architecture, Weitzman School of Design.*

state-of-the-art facilities together offer an alternative to established models of design research that support diverse, specialized career trajectories.

## Design-Research Models

The term "design research" is loosely defined in architectural education. It is used to describe studio project formulation activities alongside more scholarly history and theory research in doctoral studies (Albrecht 2019). The MSD-RAS operates similarly to scientific and engineering research models employed in more technically orientated, research-led doctoral design activities such as Harvard GSD's DDes program (Harvard Graduate School of Design n.d.), ETHZ's Gramazio + Kohler Research Lab (Gramazio & Kohler n.d.), and others. However, these PhD programs prioritize individual specialization over the more collaborative research frameworks explored in post-professional Masters degrees. Cornell's MSc Matter Design Computation program (Cornell AAP n.d.), University of Stuttgart's ICD program (University of Stuttgart n.d.), and the Architectural Association's Design Research Laboratory (Steele 2006) and Emergent Technologies programs (Hensel & Menges 2006), amongst others, offer curricula that supports students working in groups on a design-research project. While extremely successful, the degree to which these models mirror interdisciplinary, collaborative work approaches found outside academia is difficult to evaluate. In such programs, courses outside of the design thesis project are typically relegated to prerequisite or adjacent activity status. The relationships these programs forge between different curricula could be expanded, suggesting there are opportunities to explore alternative educational models.

## Hyper-Integrated Curricula

Currently, the MSD-RAS program is two semesters long and includes three design studios that primarily operate through the development of experimental, fabricated prototypes. Each studio explores a different material and fabrication approach that varies over time. The studios' design briefs are also light on professional degree architectural criteria (such as planning, site, etc.), allowing courses to focus on core knowledge domains aligned with the program's specialization in robotic fabrication and computational design. The first two half-semester-length studios in the fall semester are more tightly choreographed in design and manufacturing workflows, operating as design workshops

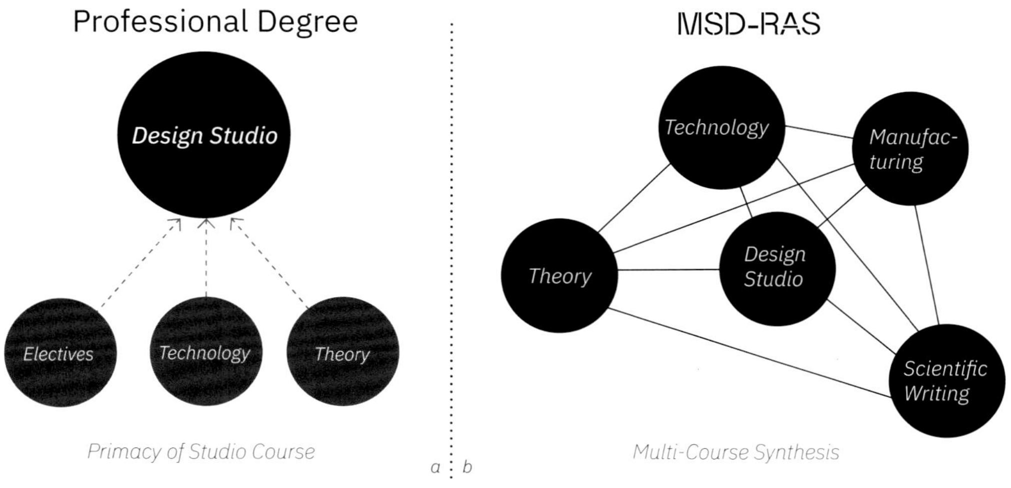


*In contrast to a professional architecture degree (a), MSD-RAS(b) courses operate in a collaborative, non-heirarchical way that builds technical and creative skills to support ambitious projects and diverse careeers.*

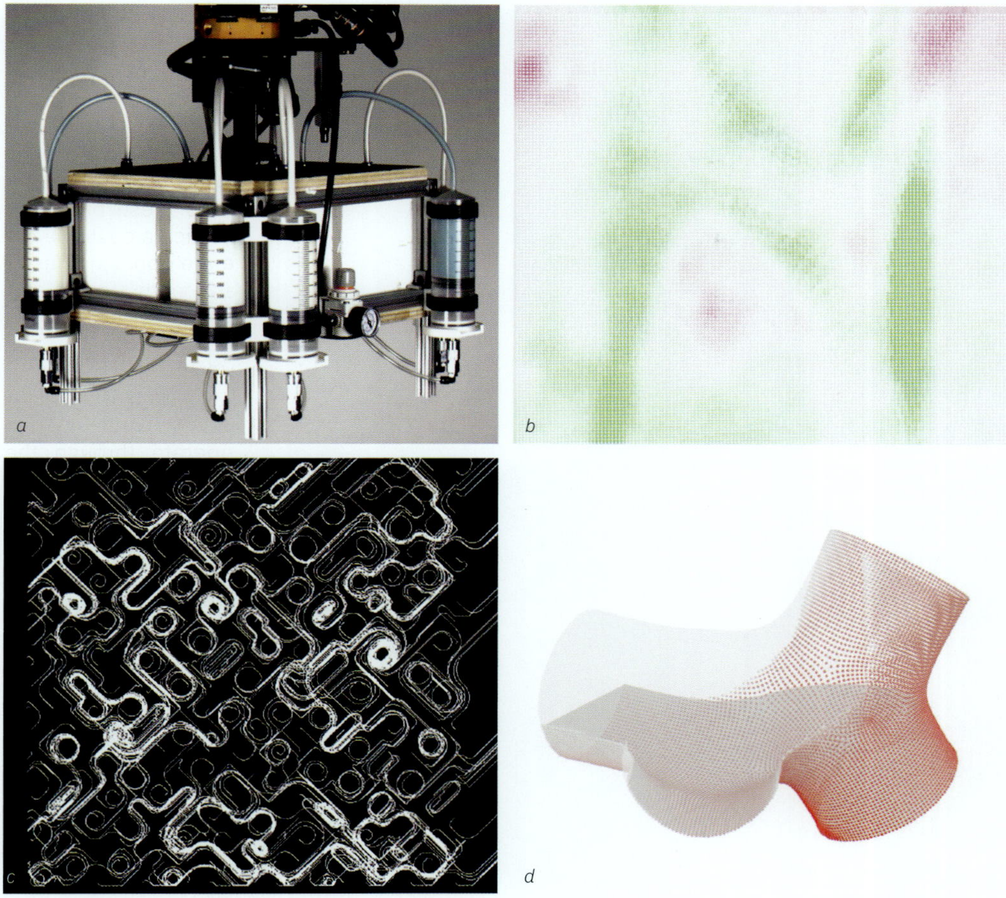

*Projects are developed across several courses in parallel to support activities such as a) custom tool design (spring project These Things Twirl), b) realtime adaptive robotic programming, (spring project Steamed Greens) c) computational generative design (spring project Filagree Façade), and d) custom developed material/fabrication simulation models. (spring project Dynamic Slip-Casting)*

that expose students to a diversity of design, computation, and manufacturing methods. Each is also paired with a concurrent half-semester-long technology course whose assignments support the design studio project and culminate in augmentations to the studio's final fabricated prototype or the design and fabrication workflow. A semester-long RAS theory course and an open elective also run through the fall semester.

In the final spring semester-long studio, a deeply integrated approach to curricula is explored. A research-led design thesis project is undertaken across four courses concurrently, including scientific research and writing, industrial robotics and tooling, advanced robot programming, and a design course that emphasizes research-led experimentation and the production of robotically fabricated physical prototypes. Given the MSD-RAS program's technical and hands-on curricula, a pedagogical structure was developed that aimed to support an integrative and collaborative approach to this shared design-research project, premised on the following:

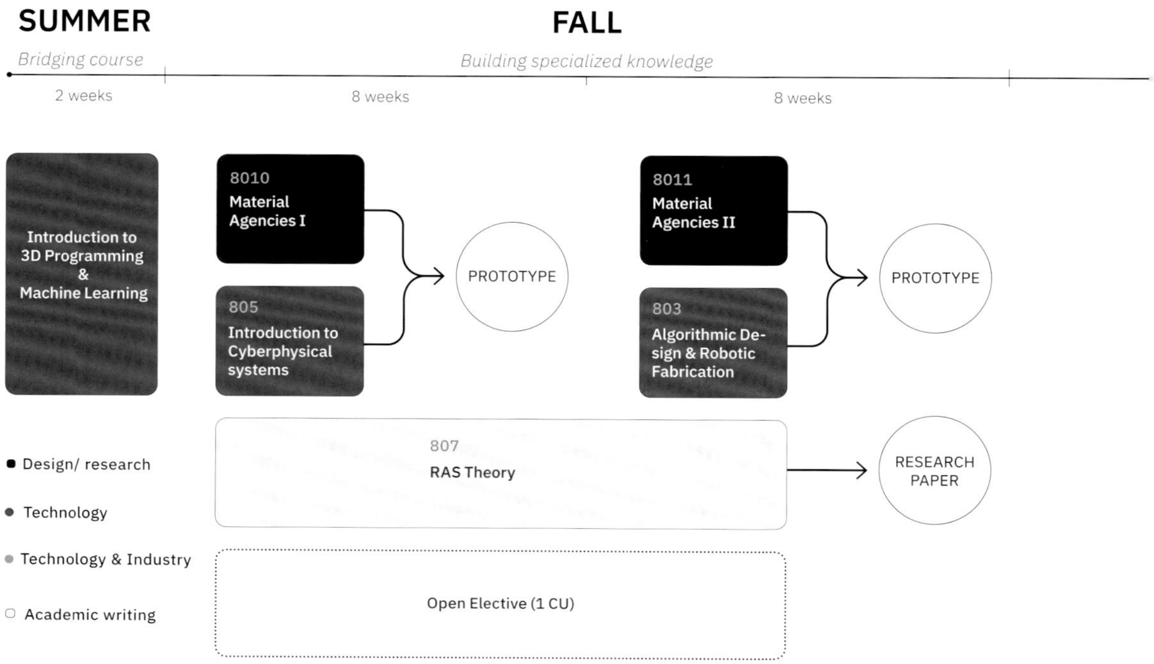


*Timeline of MSD-RAS courses that illustrates collaborative moments of review and project realization. Fall is divided into two shorter-length projects whilst spring involves a semester-length project with greater degrees of coordination and collaboration across courses.*

1. *All coursework matters equally–in contrast to the Beaux-Arts prioritization of design–and therefore all courses should form part of final thesis projects.*
2. *The implementation of (1) provides students with more agency to weigh their project towards subject areas of personal interest and aims to facilitate more diverse career opportunities post-graduation.*
3. *Application of coursework in design-fabrication work supports learning by re-enforcing attained knowledge through hands-on experience.*
4. *A speculative component in assignments encourages exploratory work beyond the pragmatics of (3).*
5. *As "design research", there is an emphasis on both aesthetically and scientifically novel work, with the value of each greatly increased when the two are intrinsically linked – conceptually and practically (e.g. a novel manufacturing method that only creates hideous designs is of little economic or cultural value while an aesthetically pleasing or provocative design that is impractical to manufacture would face many obstacles to realize).*
6. *Undertaking research on even one aspect of a design's life cycle, such as fabrication or disassembly, can be transformative and lead to novel design opportunities. Research can, therefore, be focused on a narrow set of inquiries in order to produce creative and technical mastery.*
7. *Despite (6), a project will require engaging with disparate courses' subject matter concurrently, potentially leading to a more novel, accomplished, and integrated design outcome.*
8. *Teams are more easily able to develop a project along several lines of inquiry simultaneously than individuals and can leverage all members' knowledge and skills.*
9. *To support the above, ideally, faculty from all courses participate in group thesis project formulation, review, and assessment,*

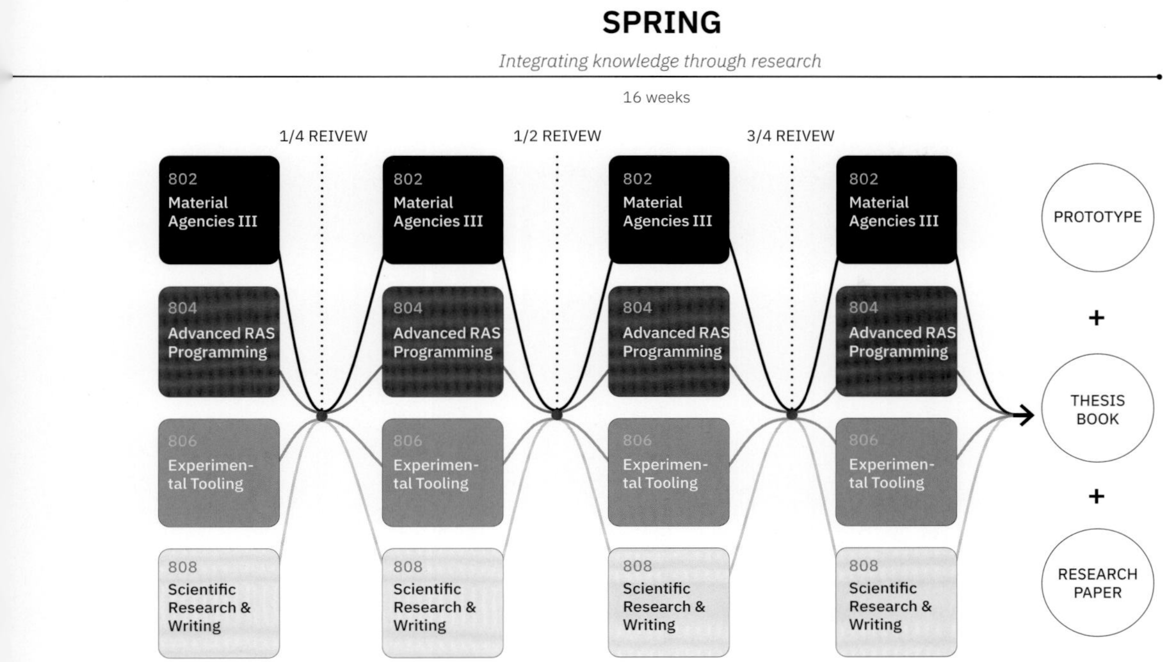


*and bring diverse, complimentary specializations to the curricula*

10. *Research-led design is best to commence with a literature review exercise and the positioning of design-research activities towards a critical and novel academic contribution. MSD-RAS teams seek not only to obtain but also to develop knowledge.*
11. *Dissemination of research-led design outcomes both contributes knowledge and connects graduates to the broader academic research community, encouraging dialogue and collective learning, and enabling students to network and build career opportunities.*

In the RAS program's first year, final semester courses ran in parallel with regular coordination meetings between instructors; however, outside of individual course schedules, students only met with the instructors all together at midterm and final reviews. In the second and third years of the program, all course instructors met students together at four intervals throughout the semester to review initial thesis proposals right through to project realization and write-up activities. This adjusted collaborative framework demonstrated noticeable improvements in the interdisciplinary nature of projects, encompassing more aspects of each contributing course's fields of knowledge.

## Design-Research Projects

In the final semester of the program, individual student teams develop a research-led design thesis project that synthesizes four courses' teaching and guidance, developing creative and critical research around industrial processes of manufacturing, robotic fabrication, and design computation. This research is primarily focused on the development of a bespoke design-to-production workflow that is explored and demonstrated through physical prototyping, computer modelling, simulation and design visualizations, and communicated in a research paper and thesis book. Successful design outcomes are intrinsic to a developed, bespoke production workflow. Individual teams' theses are direct responses to a shared architectural robotics research topic that is envisaged to change every three years. This multi-year duration allows for a collective body of knowledge to be explored that builds on previous years of experience. Students are able to critically learn from preceding RAS work and to consider it part of the state-of-the-art in which their work must be differentiated from. There are also cases where published RAS project research is able to cite previous RAS papers, rendering this form of knowledge development transparent and highlighting the program's ability to foster continuous development.

Working in teams, students commence with a literature review to gain an understanding of state-of-the-art academic research and industrial practices before identifying a design-research topic that aligns with a specific manufacturing process or processes. Physical tests are then undertaken to establish possible avenues for augmenting the manufacturing method through a robotic process. Each team also defines specific design considerations (e.g., functional, environmental, material, and ornamental agencies), and attempts to resolve these within a design approach that emerges through the development of a novel means of robotic production. A design methodology is therefore developed that integrates aspects of the manufacturing process from the outset, alleviating the need for industry practices such as geometrical rationalization or value-engineering exercises.

Final-semester RAS projects aim to develop novel robotic manufacturing approaches and corresponding design methodologies that are also then applied to a multi-part prototypical assemblage. This is no small feat as the production of the prototype takes place within an extremely short time frame towards the end of the semester with limited time to learn from manufacturing errors to refine production methods or remake problematic parts. As will be apparent in projects developed during the first three-year research agenda documented in this book, these research projects are of high quality and interest and embody both scientific rigor and aesthetic curiosity.

Undertaken within a 16 week semester from inception to completion, each design-research project is demonstrated through a robotically fabricated prototypical assemblage, a research paper, project thesis book, video, and slide presentation. The book documents a team's literature review, research experiments, production process, and corresponding speculative design proposal, embodying the entirety of the group's work across four courses during the semester, while the paper quite naturally, has a narrow focus on the most novel aspects of the research. This diverse media emphasizes different aspects of the work and clearly articulates the rigor and expertise students develop and demonstrate in their research. It also supports students launching into a diversity of exciting career.

Unlike the focused folio of a professional architect, MSD-RAS students graduate with a mega-folio that offers compelling insights into not only several design projects but also outstanding research and technical developments that were integral to each design outcome. Their photographic documentation of three studio course's fabricated prototypes also leaves no room for one to question the depth or truthfulness of the work —it is made apparent. Compared to a professional degree studio outcome consisting of representational media —scale-model buildings, drawings, perspective images, conceptual diagrams—MSD-RAS project work does not elude to what can be done, it explores at the frontier of what is possible and demonstrates this in tangible proof outcomes. Such speculative yet fabricated work provides RAS students with application-capable knowledge that is of immense value to academia and industry, and it equips graduates to be future leaders and shapers of tomorrow's built environment.

During the first three years of the program, seven out of twelve team's papers were accepted for peer-reviewed publications in world-leading conferences and journals. This equates to an almost 60% publishing rate. The significance of this cannot be understated.

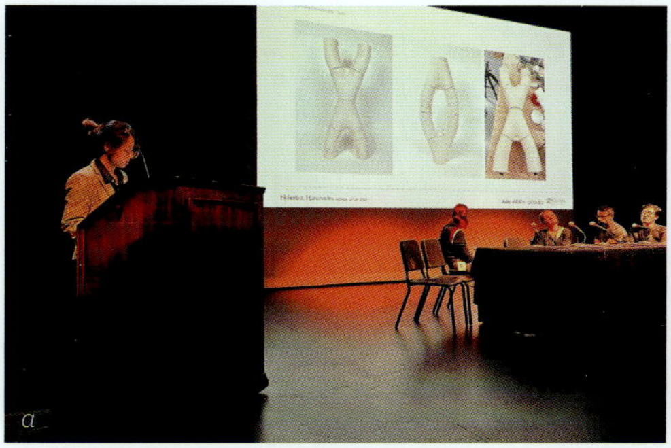

*Several MSD-RAS students have successfully published and presented their thesis projects in peer-review conference proceedings: a) ACADIA 2022, b) ACADIA 2023*

Firstly not all twelve projects submitted papers to conferences, and secondly, these MSD-RAS students' 16-week projects were competitively selected adjacent to fourth-year PhD research and tenured professors' multi-million dollar funded research to present in conferences where the acceptance rate of any paper is typically less than 30%. Beyond competitive, this speaks to the type of knowledge being developed and disseminated by MSD-RAS students. These scientific research papers cover experimentation, data collection, and provide detailed instructions that inform others how to replicate their findings, contributing practical open-source knowledge. Graduates presenting their work are welcomed into a large academic and industry community that is deeply interested in their work, providing a platform for students to network and develop their next career steps post-graduation.

## High-Calibre and Collaborative Faculty

Perhaps more critical than the pedagogical structure of the MSD-RAS program is the calibre of the MSD-RAS faculty, each of whom was mentioned in the introduction, who are all leaders in their respective and diverse fields of expertise. Beyond their knowledge and experience, their passion for teaching, also has a significant impact on student levels of motivation and achievement.

As was apparent in the introduction, each MSD-RAS instructor brings a large body of knowledge to the program, and several have authored books on subjects extremely relevant to MSD-RAS curricula. To illustrate this point, Nathan King's co-authored *Ceramic Material Systems: in Architecture and Interior Design* (Bechthold et al. 2015) became a valuable resource during the first three-year research agenda into architectural ceramics, while our new research focus on architectural thermoplastics commences from a context extremely well-researched in Billie Faircloth's *Plastics Now: On Architecture's Relationship to a Continuously Emerging Material* (Faircloth 2015). With multi-year research agendas exploring façade screens and now building envelopes, Jeffrey Anderson's co-authored The *Ecologies of the Building Envelope: A Material History and Theory of Architectural Surfaces* (Zaera-Polo et al. 2021) becomes extremely relevant. Complimenting these, the author's book *Behavioural Production: Semi-Autonomous Approaches to Architectural Design, Robotic Fabrication and Collective Robotic Construction* (Stuart-Smith 2024), provides a conceptual framework for generative computational design and robotic fabrication. These publications are also matched by seminal examples of robotically fabricated architecture, such as Andrew Saunders Deep Relief Project (Saunders et al. 2022), Nathan King, Autodesk and DAR's Smart Bridge (Storey 2023), or the author's demonstration of Aerial Additive Manufacturing (Zhang et al. 2022). Some of these are featured in the Extra-Curricula section of this book.

It is rare to have so much in-depth knowledge embodied within a new program's faculty, or for such high-calibre faculty to be engaged in such an ambitious collaborative form of

*The ARI Robotics Lab is packed with state-of-the-art robots, end-of-arm tools and infrastructure to support diverse research activities. The facility is constantly evolving to support developments in the MSD-RAS curricula.*

design-research education. While other post-professional programs can have several layers of management between experienced faculty and students, in the MSD-RAS program, these industry-leading faculty are directly interfacing and working with students on a weekly basis. The engagement of the MSD-RAS faculty goes further still. In the final semester, faculty participation in regular coordination meetings and quarterly reviews provides a forum in which hive-mind thinking can drive strategic decision-making across curriculum boundaries. This dialogue also builds a critical and creative consensus around the program's multi-year research agendas and opens up a constantly evolving shared exploratory space that is inducive to progressive work. In addition to the faculty and pedagogy, the success of the MSD-RAS educational model is also directly tied to the program's new robotic fabrication facilities and their co-development in lockstep with the program.

## Rethinking Curricula and Infrastructure

The program's capacity to do frontier work in robotic fabrication is quite obviously dependent on access to facilities and equipment that can support this type of work. A student's knowledge and skills development in industrial methods of robotic fabrication is also directly related to their training with specific facilities and personnel. The success of the MSD-RAS program is contingent on this tight intertwining of these two disparate arenas — academic curricula, faculty, and programming on one hand and robotics lab operational and equipment procurement decisions on the other. By necessity, in a program such as the MSD-RAS, curricula are constrained by the capacity of a robotics lab to support particular materials and means of manufacture or robot control. New equipment, safety protocols, and operational logistics must be planned and implemented in advance to support new research trajectories. Recognizing this, the MSD-RAS program director also co-supervises the ARI Robotics lab, ensuring strategic procurement and hiring decisions, while day-to-day operations in the facilities provide MSD-RAS students with the capacity to realize progressive work. The ARI Robotics Lab is fortunate to have exceptional full-time managers who supervise part-time staff, support students' work, and manage operational, maintenance, and lab advancement activities.

This highly coordinated operational and technological infrastructure is managed in such a way as to provide students with a maximum amount of support while not overdoing any handholding. In the MSD-RAS, students are taught how to develop and manage robotic fabrication processes in their entirety. This in-depth knowledge builds independence, empowering students to be able to continue their work throughout their careers. No better evidence of this exists than the number of graduates who have served as robotics lab staff post-graduation. The knowledge gained in the MSD-RAS program is sufficiently deep and expansive to enable graduates to work effectively as managers or assistants in the lab post-graduation, capable of training the next generation of students.

In MSD-RAS, student work is also supported by a materials budget. Students' material needs are supplied and funded by the program, alleviating what would be considerable financial and time pressure on students. This provides some financial equality to students' activities and removes an otherwise real barrier to progressive research.

### State-of-the-Art Facilities

Evaluating a program's educational offering based on its equipment is difficult. As might be ascertained from the operational and planning activities mentioned above, there is a lot more at stake. It is critical to consider students' access, knowledge, and support for using such equipment, or their ability to initiate and execute work with minimal impedance. While many architecture programs have robots bolted to floors, at Penn, a comprehensive approach was taken towards strategic planning of robot platforms, tooling, control, safety, and environmental infrastructure. The result is an architectural robotic fabrication facility that is world-leading in its work-cell integration and extensibility, which radically impacts what students can do in a robotics space.

This unique approach to facility planning arose from observations of the operational differences between robotic fabrication in factories compared to academic coursework requirements. In a factory, robots typically perform a small number of repetitive tasks in a fixed work cell that has fixed requirements in terms of tooling and environmental management. This approach is in direct contrast to the needs of an academic program where material fabrication activities and tooling might vary regularly throughout a single day and must also accommodate larger changes to lab infrastructure over time. To maximize efficiency and return on such equipment in an academic setting, it is also imperative that several tools, materials, and work-cell spatial arrangements can operate from the same space at different times with minimal change-over time overhead. To support more flexible academic research, the ARI Robotics Lab incorporates the following:

- *Rapid Task-Changing:* Our robot work cells are equipped with a large number of robot tools that can be quickly connected to a robot using an automatic tool-changing system that also reconnects the tools I/O to the robot (such as power, data, air, or sensor connections). This not only saves

*MSD-RAS Material Agencies design studios expose students to diverse materials and manufacturing processes. a) a visit to a timber supplier, b) lecture from Stella Dourtme on SOM/Zaha Hadid façade design, fabrication, and detailing.*

backbreaking effort maneuvering 50 kilogram tools into position or the need for human hands to reestablish 480-volt power connections, but it also speeds up changeovers between students and robot tasks. More fundamentally, it also enables manufacturing to incorporate several tools within one production process.

- *Extensibility in hardware:* Students develop their own robot tools in the MSD-RAS program. An infrastructure is in place to enable these custom tools to be quickly plugged into a robot's tool racks and I/O to enable sensors, power, air, and signals to be integrated within robot programs. Swapping out different custom tools is extremely quick.
- *Extensibility in software:* ARI Robotics Lab robots can be controlled by industry-standard programming methods (specific to each robot's brand – e.g., ABB rapid code), along with additional programming approaches that enable real-time control or live adaptation to sensor feedback throughout fabrication. Students are thus able to hack the robots to perform specifically curated routines that are aligned to bespoke control strategies.
- *Safety and monitoring systems:* Dangerous industrial manufacturing procedures, such as 7-axis milling are undertaken alongside more user-friendly activities, such as clay additive manufacturing. A custom monitoring and control system enforces different safety procedures appropriate for each respective tooling or material activity.

## Agile, Integrated, and Evolving

The MSD-RAS program explores architectural design through integrated approaches to robotic fabrication and computational design. As discussed in this book's essay "Towards Integrated Design & Production," any practical or logistical consideration in the fabrication, transportation, or assembly of architecture can be conceived to contribute creatively to a design outcome. The design of a robot tool, for instance, can have aesthetic impacts on a fabricated outcome if strategized to do so. This expansive stance on the media and means of design quite naturally applies also to the design of the program itself.

*Assembly of 8010 Material Agencies fabricated prototype. MSD-RAS Projects are a collaborative undertaking.*

MSD-RAS' pedagogical structure, its curricula, faculty, and state-of-the-art facilities were all conceived to operate together, to support novel design research that is greater than the sum of its parts. Metaphorically, this interrelated grouping is less machinic and more ecological in nature. Relationships between each of the MSD-RAS' diverse ingredients (faculty, equipment, materials, operational logistics, research agendas, etc.) can be redefined over time. It is hoped this integrative approach to collaborative learning fosters novel design research and, in parallel, empowers agile graduates, well-versed in the knowledge and skills of today to positively impact architecture's challenging yet exciting future.

*– Some parts of this essay were first published by the author in "Towards Technically Progressive and Speculative Design Education: Penn MSD-RAS's Collaborative and Interdisciplinary Pedagogy," presented at the 2023 ACSA/EAAE Teachers Conference: Educating the Cosmopolitan Architect, held in Rekyavik, Iceland, June 22-24 (Stuart-Smith 2023).*

## References


Albrecht, L. (2019). "Inquiry-Based Learning in Architecture." In *Inquiry-Based Learning--Undergraduate Research: The German Multidisciplinary Experience*, edited by H.A. Mieg, 249-258. New York: Springer International Publishing, 2019.

Bair, K., F. Gandhi, C. Griffin, S. Kim, M. Neveu, H. Rashed-Ali, M. Uihlein, K. Tsubaki, and E. Ellis Wayne. "ASCA White Paper on Architectural Education Research and STEM." Association of Collegiate Schools of Architecture, 2018.

Bechthold, M., A. Kane, and N. King. Ceramic Material Systems: In *Architecture and Interior Design*. Basel: Birkhäuser, 2015.

Cornell AAP. (n.d.). "Master of Science, Matter Design Computation." Cornell. Accessed on October 10, 2022. https://aap.cornell.edu/academics/architecture/graduate/mdc

Cunningham, A. "The Genesis of Architectural Education." Studies in *Higher Education* 4, no. 2 (1979): 131-142. https://doi.org/10.1080/03075077912331376917

Cunningham, A. "Notes on Education and Research Around Architecture." *The Journal of Architecture* 10, no. 4 (2005): 415-441. https://doi.org/10.1080/13602360500285542

Faircloth, B. *Plastics Now: On Architecture's Relationship to a Continuously Emerging Material*. London: Routledge, 2015.

Gramazio, F., and M. Kohler, "Gramazio Kohler Research ETHZ." Gramazio Kohler. Accessed on October 10, 2022. https://gramaziokohler.arch.ethz.ch/web/news/e/actualNews/0/0/0.html

Gropius, W., and P.M. Shand. *The New Architecture and The Bauhaus*. Cambridge: MIT Press, 1965.

Harvard Graduate School of Design. (n.d.). "Doctor of Design Program." Harvard. Accessed on October 10, 2022. https://www.gsd.harvard.edu/doctoral-programs/doctor-of-design/

Hensel, M., and A. Menges. "Material and Digital Design Synthesis." *Architectural Design* 76, no. 2 (2006): https://doi.org/10.1002/ad.244

NAAB. "Accreditation." 2023. https://www.naab.org/accreditation/

Ng, R. "Breaking the Chains: Beyond the Beaux-Arts Tradition of Architectural Education in the United States." In *Rethinking the Crit: New Pedagogies in Design Education*, edited by P. Flynn, M. O'Connor, M. Price, and M. Dunn, 174-182. London: Routledge, 2022.

Saunders, A., R. Studebaker, and C. Eileen Moriarty. "Deep Relief." In *Hybrids and Haecceities: Proceedings of the 42nd Annual Conference of the Association for Computer Aided Design in Architecture*, edited by M. Akbarzadeh, D. Aviv, H. Jamelle. and R. Stuart Smith, 652-661. La Vergne: IngramSpark, 2023.

Schwab, K. *The Fourth Industrial Revolution*. New York: Crown Publishing Group, 2017.

Steele, B. "The AADRL: Design, Collaboration, and Convergence." *Architectural Design* 76, no. 5 (2006): 58-63. https://doi.org/10.1002/ad.323

Storey, P. "Engineering a Bridge that Designs and Builds Itself." Autodesk Research. July 2, 2023. https://www.research.autodesk.com/blog/engineering-a-bridge-that-designs-and-builds-itself/

Stuart-Smith, R. "Towards Technically Progressive & Speculative Design Education: The Collaborative & Interdisciplinary Pedagogy." 2023 ACSA/EAAE Teachers Conference: Educating the Cosmopolitan Architect.

Stuart-Smith, R. *Behavioural Production: Semi-Autonomous Approaches to Architectural Design, Robotic Fabrication, and Collective Robotic Construction*. London: Routledge, 2024.

University of Stuttgart. (n.d.). "Institute for Computational Design and Construction." Accessed on October 10, 2022 from https://www.icd.uni-stuttgart.de/

Zaera-Polo, A. and J. Anderson. *The Ecologies of the Building Envelope: A Material History and Theory of Architectural Surfaces*. New York: Actar Publishers, 2021.

Zhang, Ketao, Pisak Chermprayong, Feng Xiao, Dimos Tzoumanikas, Barrie Dams, Sebastian Kay, Basaran Bahadir Kocer, Alec Burns, Lachlan Orr, Talib Alhinai, Christopher Choi, Durgesh Dattatray Darekar, Wenbin Li, Steven Hirschmann, Valentina Soana, Shamsiah Awang Ngah,Clément Grillot, Sina Sareh, Ashutosh Choubey, Laura Margheri, Vijay M Pawar, Richard J Ball,Chris Williams, Paul Shepherd, Stefan Leutenegger, Robert Stuart-Smith and Mirko Kovac. "Aerial Additive Manufacturing with Multiple Autonomous Robots." Nature (italicized) 609, no. 7928 (2022): 709-717. doi:10.1038/s41586-022-04988-4.

# BRIDGING

## Introductory Course

Summer

# SUMMER COURSE

**The MSD-RAS program commences with a two-week introductory course prior to the start of the fall semester. The course runs on campus during business hours, Monday through Friday, and aims to help incoming students to transition smoothly into the MSD-RAS curricula and to provide them with the necessary conceptual and technical skills for advanced architectural design research in robotics and autonomous systems.**

**800 Summer Course**
(Coding/3D Modeling/Robot Programming)

Instructors:
Ezio Blasetti (2020-2021)
Patrick Danahy (2022-)

Contemporary Architectural Design has been radically redefined in the last couple of decades through a series of advances in algorithmic design, robotic fabrication and deep learning (AI). Computation, digital media, and fabrication have shifted fundamental methods of conception, drawing, modeling and construction in architecture. Algorithmic tools are often at the intersection of different disciplines allowing for novel domains of interdisciplinary research. This course provides a foundation for computational design with both generative and procedural (parametric) approaches to 3D design being explored in addition to a basic introduction to robot programming and machine learning. In addition to live-instruction an

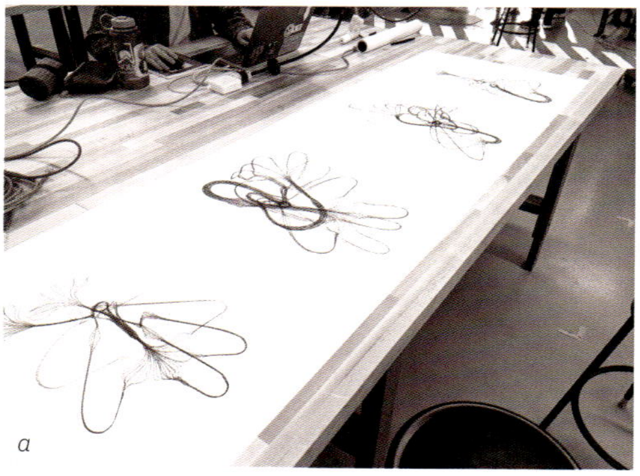

a

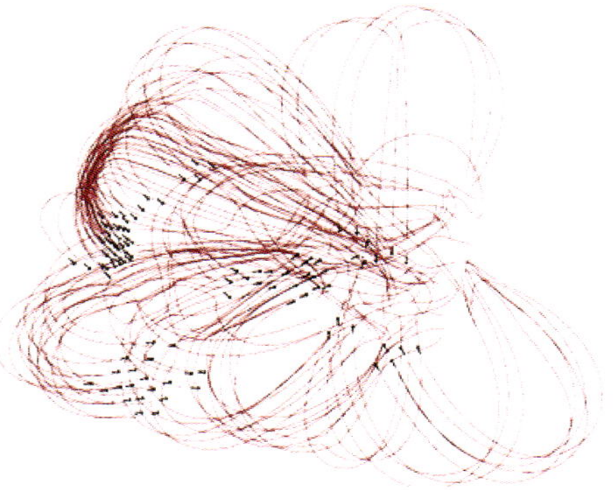

b

*a) Robot drawings, drawn by an ABB IRB120 robot. The linework was generated by b) a custom-developed multi-agent simulation. Student: Davis Dunaway.*

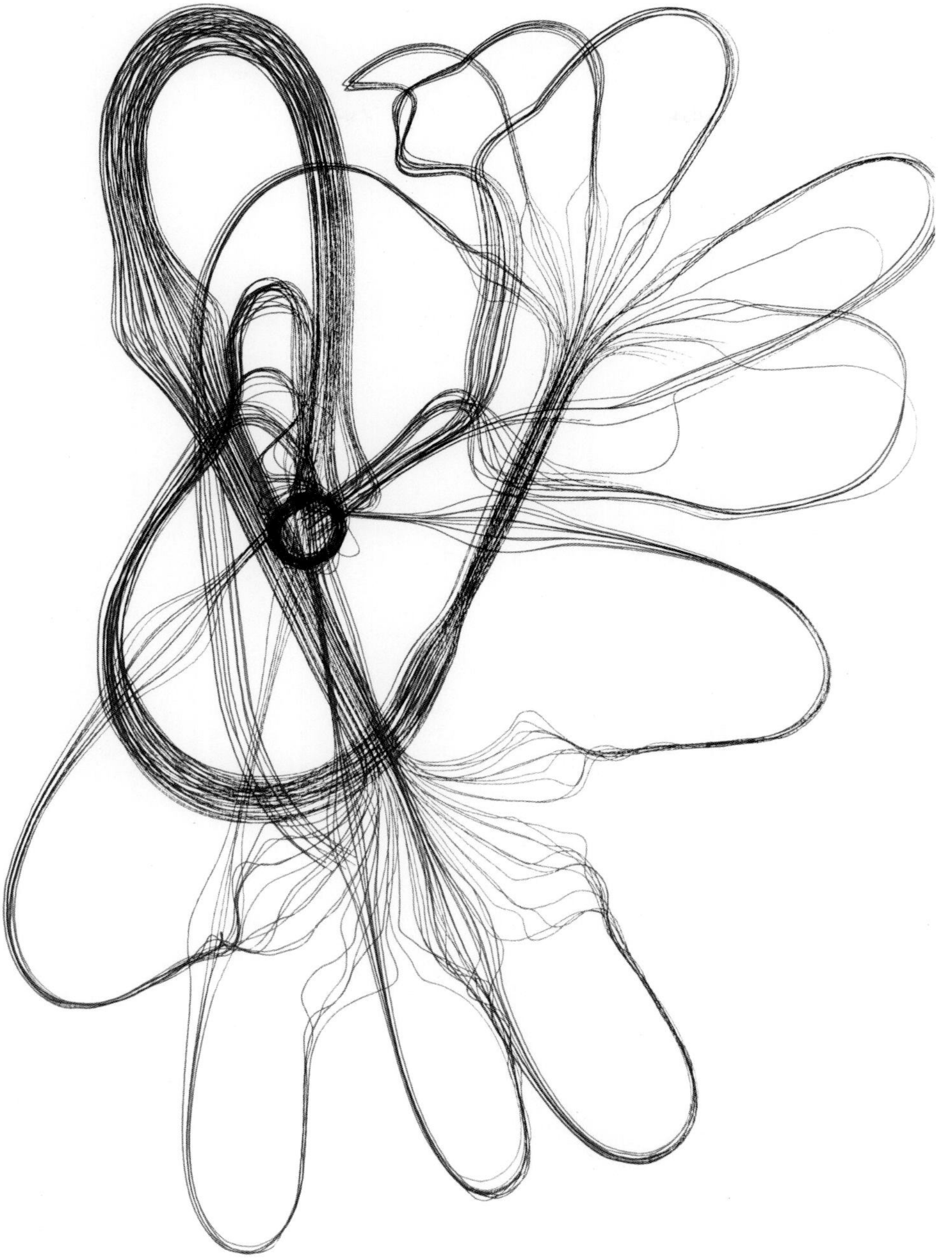

*Multi-agent generated robot drawings, drawn by an ABB IRB120 robot. Agents spatially interact with each other generating a series of motion trails which are then drawn by a robot grasping a pen. Student: Davis Dunaway.*

assignment is undertaken during the workshop that helps learned knowledge to become tacit knowledge. The specifics of the assignment and tutorials varies per year as do learning objectives. In principle, however, participants develop an introductory understanding of computational goemetry, generative design, and robot programming.

*Computation and Geometry*
Different computational (scripted) and explicit (modeled) descriptions of geometry and their corresponding spatial datatypes in the Rhino3D and Grasshopper environment are engaged.

*Generative Design*
Participants are introduced to generative design through object-oriented programming in Python in Rhino3D and in web-based platforms such as Google Colab. Emphasis is on non-linear processes such as self-organization, deep learning processes such as Deep Neural Networks for image-generation, and analysis processes such as spatial, structural, or geometrical analysis. Students develop and synthesize their own algorithms for spatial analysis and/or form generation.

*Robotic Fabrication*
Students gain a basic, initial introduction to producing and simulating a custom toolpath for any 6-Axis Industrial Robot and may undertake a simple simulated or physical robotic fabrication exercise.

Through this introductory course, MSD-RAS students receive a comprehensive introduction into critical elements in the digital workflow of MSD-RAS curricula and projects that are expanded on in fall and spring semesters. Students develop sufficient computational literacy and geometric proficiency to support their subsequent design and research activities related to robotic fabrication for architectural design. The course has been strategized to ensure students from diverse backgrounds gain an equal footing to commence their RAS studies whilst students with some prior experience in the workflows of the program are also stretched into new territories.

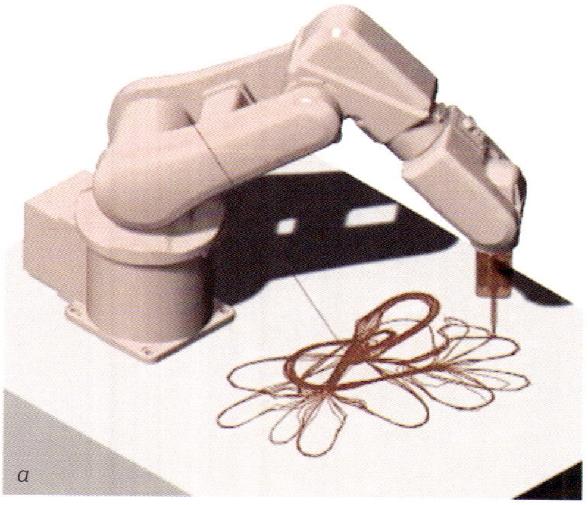
a

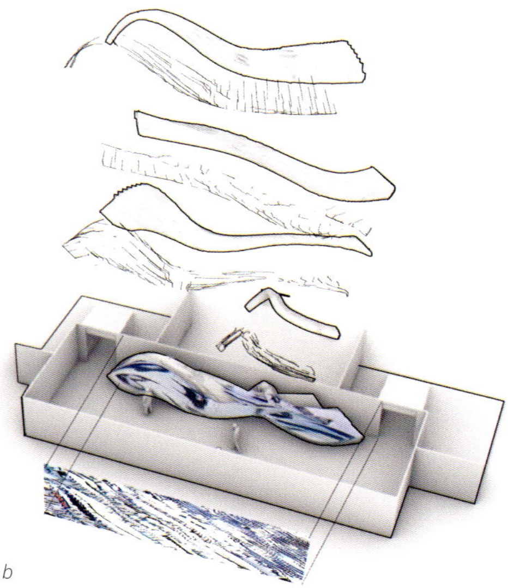
b

*a) Simulation of a robot manufacturing program, b) 3D translation of style transfer deep-learning into figure, relief and color, c-e) Algorithmically generated field conditions, deep learning image generation: f) content image from 3d model, g) style image, h) style transfer result, i-k) 3D wall instalation designs. Students: a) Davis Dunaway, b) Joey Luo, Sihan Li, Shane Su, c-e) Siyu Dong, f-j)Franklin Wu, Amber Chen, k)Nicholas Houser, Bentian Wang, Jeffrey Liao.*

c

d

e

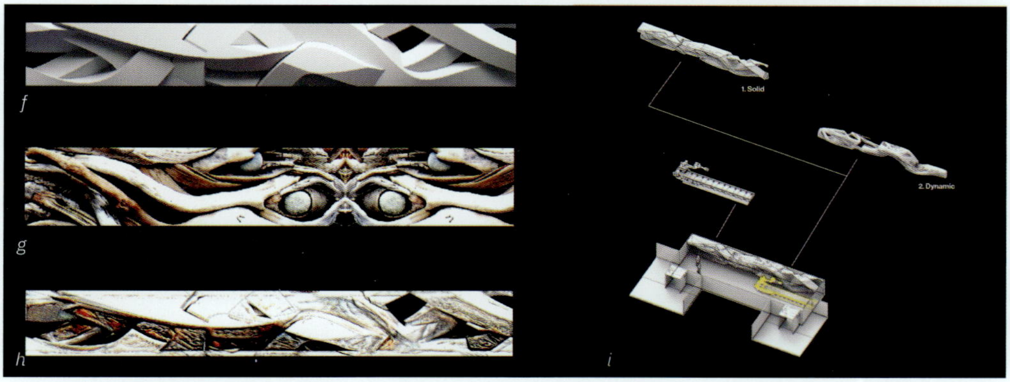
f
g
h
1. Solid
2. Dynamic
i

j

k

# BUILDING

## Specialized knowledge

*Fall Semester*

## Fall Semester

*The MSD-RAS runs over two semesters. In the fall semester, students undertake a semester-long RAS theory course, an elective course, and two eight-week design projects that are centered around the development of full-scale, robotically manufactured prototypes. Each of these projects explores different material, manufacturing and design approaches, that are applied to a different project brief. Each project is undertaken across two eight-week courses, including a Material Agencies: Robotics and Design Lab course and a technology course, totaling four courses. Material Agencies: Robotics and Design Lab I and II are similar to an architectural design studio course except that they operate primarily from Weitzman School's ARI Robotics Lab and emphasize designs being developed through iterative experimental prototyping. Material Agencies seeks to synthesize robotic manufacturing and design methodologies in projects that address a design brief. The brief is relatively simple, alleviating architectural concerns such as program in order to focus on the core design research associated with the course. Each technology course runs in parallel to one of the Material Agencies courses, teaching complementary knowledge and skills in the development of cyber-physical systems, robotics, and generative computational design. The specifics of each course are included in the book's appendix, while the subject matter of the project briefs shared by the Material Agencies and technology courses are introduced ahead of student projects presented in the following pages for the two half-semester fall projects. No student work from the RAS Theory is presented in the book due to space limitations; however, information on the course is provided on the next spread.*

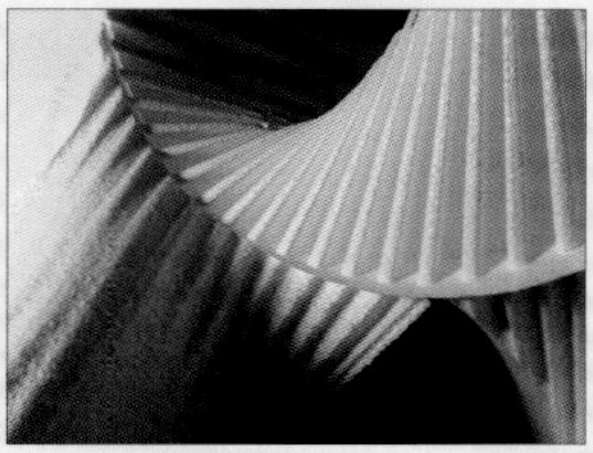

## *Fall Section I*

**8010: Material Agencies Robotics and Design Lab I**
Instructor: Andrew Saunders

**803 Introduction to Cyber-Physical Systems**
Instructors: Jeffrey Anderson(2022—), Ezio Blasetti(2020-21)

## *Fall Section II*

**8011: Material Agencies Robotics and Design Lab II**
Instructor: Ezio Blasetti

**805 Introduction to Algorithmic Design**
Instructor: Ezio Blasetti(2020-22), Patrick Danahy(2023—)

*Architecture in the Age of Thinking Machines*

**807: RAS Theory**
Instructor: Evangelos Kostioris

**Fall Elective Course**
Various Instructors

## RAS Theory: Architecture in the Age of Thinking Machines

Instructor: Evangelos Kotsioris

Robotics, autonomous systems, advanced methods of fabrication, and AI — are all terms that seem to have become commonplace among architects. Yet their origins, evolution, and philosophical underpinnings often remain obscure. This seminar is intended as a historical and theoretical introduction to a series of key concepts that are central to the MSD Robotics and Autonomous Systems program. Using contemporary discourse and concerns as a jumping-off point, the course (re)constructs an archaeology of interdisciplinary knowledge central to design computation, machine intelligence, and automated fabrication. The intellectual origins of these areas, and their profound consequences for architecture, are traced back to the fields of technology, communication theory, cybernetics, systems theory, developmental biology, robotics, network theory, aesthetics, psychology, and artificial intelligence, among others. These topics also influence socio-political themes around modes of production and automation, and philosophical discourses such as posthumanism, the Anthroprocene, vital materialism, object-oriented ontology (OOO), and digital materiality, amongst others.

The course runs as a weekly seminar with readings, student presentations, and guest lectures throughout the semester by Molly Wright Steenson, Anna-Maria Meister, Lydia Kallipoliti, Theodora Vardouli, David Theodore and others. A written theory paper is the primary submission for the course. A combination of historical sources and contemporary texts provides students with a common lexicon and set of references that allow them to articulate and critically assess their own conceptual and material explorations in robust ways. Particular emphasis is placed on the philosophical, socio-political, environmental, and ethical extensions of existing and emerging digital technologies, including the adverse effects of profiled surveillance, automated decision-making, and algorithmic bias. Weekly readings provide a basis for discussions but also serve as an interpretative lenses for the analysis of case study projects. Supplementary sources also enable students to deepen their literacy on specific topics beyond the completion of the seminar.

## Fall Elective Courses

In fall, students undertake one semester-long elective. Electives offered vary per year. Electives in computer science, fine arts, and business were undertaken together with the following electives offered in the Weitzman School of Design listed below:

**724-001: Technology in Design: Immersive Kinematics/Physical Computing: Body As Site**
Instructor: Simon Kim

**725-401: Design Thinking**
Instructor: Sarah Rottenberg

**731-001: Experiments in Structure**
Instructor: Mohamad Al Khayer

**732-002: Technology Designated Elective: Material and Structural Intelligence**
Instructors: Sameer Kumar & Mark Nicol

**732□003: Technology Designated Elective:** Geometric Structural Design
Instructor: Masoud Akbarzadeh

**732-004: Technology Designated Elective: Matter and Energy Design**
Instructor: Franca Trubiano

**732-005: Matter, Making,And Testing: Designing With Next Generation Precast Concrete**
Instructor: Richard Garber

**732-006: Heavy Architecture**
Instructor: Philip Ryan

**7325: Inquiry into Biomaterial Architectures**
Instructor: Laia Mogas-Soldevila

**743-001: Form and Algorithm**
Instructor: Ezio Blasetti

**768-402 Real Estate Development**
Instructor: Alan Feldman

**765-001: Project Management**
Instructor: Charles Capaldi

**LARP 740-002: Topics in Digital Media: Game Design Toward Posthuman Relational Aesthetics**
Instructors: James Billingsley & Patrick Danahy

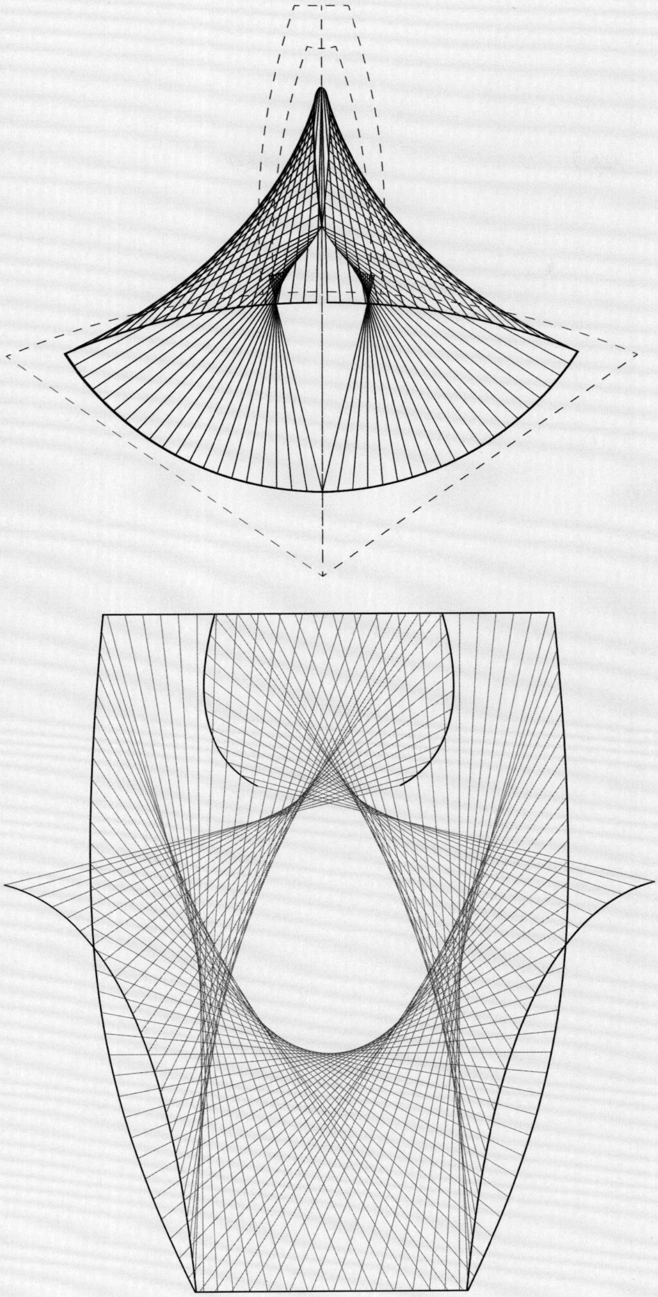

*Above: Ruled-surface wireframes of robot hot-wire cutting toolpaths for volumetric studies of Pevsner/ Gabo sculptures. Next page: Robot hot-wire cut EPS foam fabricated ceiling prototype. Students: Christine Khouri Sader, Vicny Yang, and Ryan Liao.*

# FALL SECTION I

In the first half of the fall semester students designed and produced a full-scale partial ceiling prototype. The ceiling was designed through integrated approaches to geometry: generative design using machine learning and robot hot-wire cut fabrication. The prototype also operated as a cyber-physical system, incorporating sensors and various means of producing kinetic behavior, lighting, or sound effects. The project was undertaken across two courses – a design studio and a technology course.

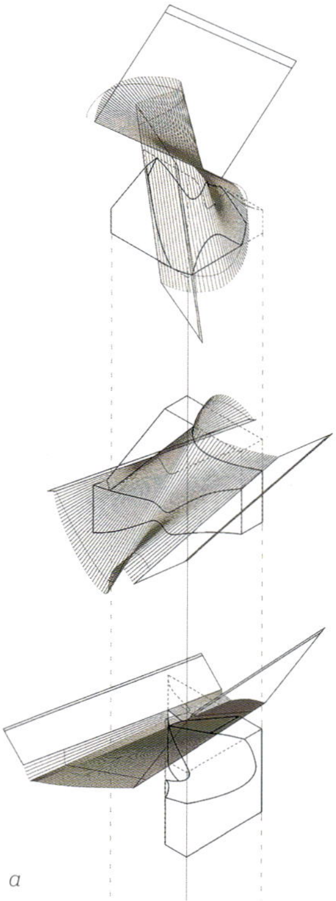

*a*

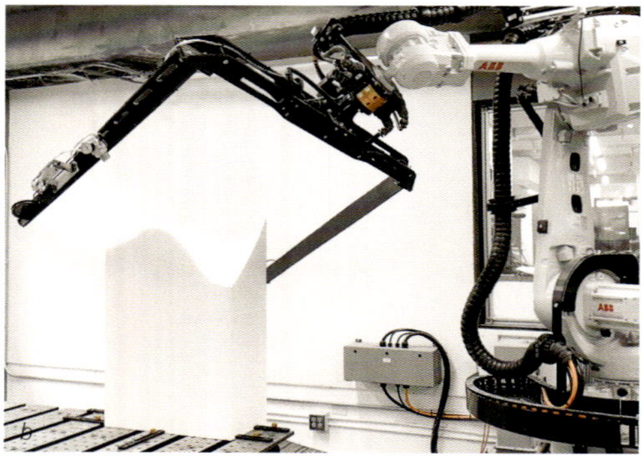

*a) Ruled-surface robot hot-wire cutting toolpaths for volumetric studies of Pevsner/Gabo sculptures, b) robot hot-wire cutting of EPS foam, and c) student presentation including fabricated ceiling prototype.*

## 8010 Material Agencies: Robotics and Design Lab I

Instructor: Andrew Saunders
TAs: Riley Studebaker, Claire Moriarty, Matthew White, Caleb Ehly, and Yujie Li.

This half-semester introductory studio to the MSD-RAS program, explores robot hot-wire cutting of EPS foam for the manufacturing of a deep-relief ceiling prototype. Working in the Weitzman School of Design's ARI Robotics Lab, students learn to program ABB IRB4600-60 six-axes industrial robots that are coupled with a hot-wire cutter end-effector. This relatively simple manufacturing task geometrically constrains designs to ruled-surface geometries that are able to be cut through varied degrees of rotation of the straight line of the tool's hot wire. Designs, thus, must be developed to be intrinsically related to the manufacturing method to avoid the need for post-design geometric rationalization.

The course commenced with an analysis of ruled-surface sculptures created by the Constructivist brothers who published the "Realistic Manifesto," Naum Gabo and Neemia Pevzner. Students recreated these sculptures as both geometry and tool paths for robotic fabrication, introducing a bespoke set of micro-motions to the tool path to create unique tooled micro-relief patterns in the fabricated parts that visually reinforced the geometry's character whilst embodying characteristics made possible due to the robot manufacturing process. Speculative architectural designs for a new ceiling-scape in one of the large galleries in Meyerson Hall were developed that extended the fabricated studies through Deep Learning, a recent and rapidly developing branch of artificial intelligence. Students identified local geometric signatures from their fabricated works and developed methods to deploy these features within a larger ceiling-scape proposal using Deep Neural Networks such as Convolutional Neural Networks (CNNs) and Generative Adversarial Networks (GANs). As a critique to the Fordist tendencies of the existing standardized dropped ceiling, each proposal explored the autonomous latent space operating between the geometric rules learned in the first

*d) Interior perspective of ceilingscape. Students: Xiangguo Cui, Siyu Dong, Davis Dunaway, and Jingjing Yan.*

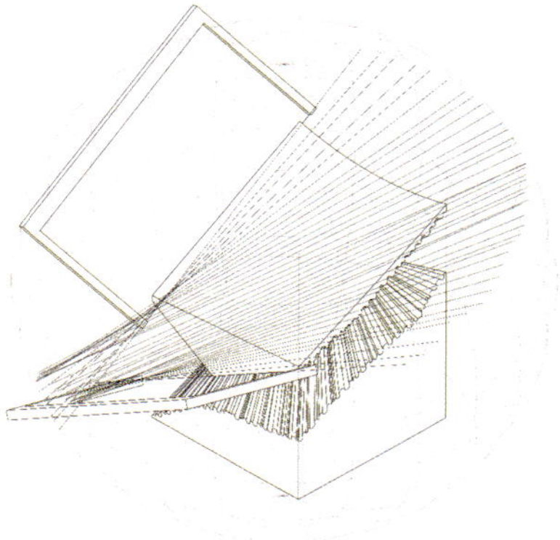

a

b

phase of analysis and other 3D modelled and sourced image content. The new ceilings aimed to embody completely unfamiliar part-to-whole relationships, resisting familiar tropes such as Vitruvian, Fordist, Collage or Parametric.

At the end of the course, each team presented a design proposal for the entire 900 sf gallery space ceiling together with a 1:1 scale robotically fabricated partial mock-up. The final review took place in the actual gallery space, with mock-ups installed on the gallery's ceiling providing a clear indication of their scale and impact in the real space. It was exciting to see this speculative work, one century after the "Realistic Manifesto," exploring a new regime of automated fabrication and autonomous systems.

## 805 Introduction to Cyber-Physical Systems

Instructors: Jeffrey Anderson (2022–), Ezio Blasetti (2020-21).
TAs: Claire Moriarty and Sophia O'Neil.

This course introduces students to electrical circuits, microcontrollers and their programming, and focuses on the development of interactive hardware systems that respond to environmental stimuli. The work is directed towards the generation of novel physical and digital architectural effects by integrating these responsive systems into robotically fabricated models and prototypes. The final deliverable for the course has students apply what they learned throughout the semester within their final 8010 Material Agencies design projects to create an interactive lighting element that becomes embedded within their fabricated architectural ceiling prototypes.

In the first part of the semester, students undertake several exercises which introduce them to programming in C++ and the Arduino platform. Students are challenged to create both a software component and a hardware component in each exercise, providing them with practical experience in working with both programming concepts and the physical material constraints of sensors, microcontrollers, and other hardware. During this time, students learn practical concepts such as control, data logging, feedback, and networking and develop craft in digital fabrication, integrating these circuits into designed objects which are 3D printed, CNC milled, or laser cut. In the second part of the semester, students work in groups to design and manufacture an interactive lighting element that they integrate into their robotically fabricated ceiling prototype. Employing Neopixel addressable LED lights, students

*a) Ruled-surface robot hot-wire cutting toolpaths for volumetric studies of Pevsner/Gabo sculptures, b)b) proximity sensor-activated lighting schemes operating in fabricated ceiling prototype, c) Machine Learning (ML) Style Transfer input images created from 3D modelled robot hot-wire cut study models, d) ML Style Transfer generated images informed by ruled-surface fabricated prototypes, e) 3D modeled ceilingscapes developed from ML Style Transfer outputs, and f) exhibition-format student final review with study*

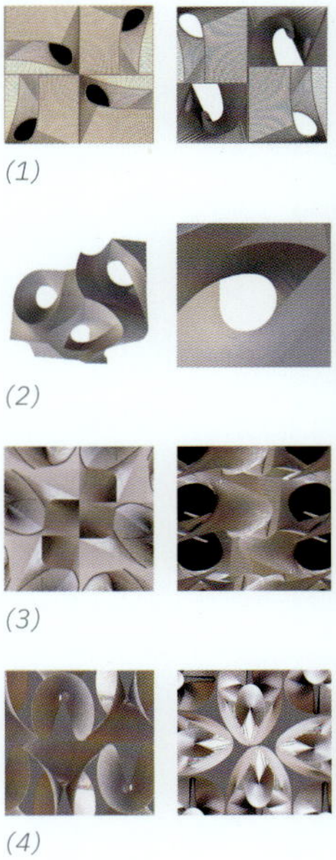


c

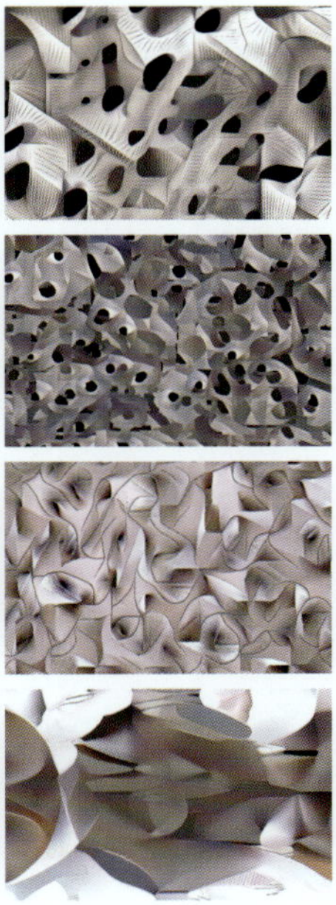

d

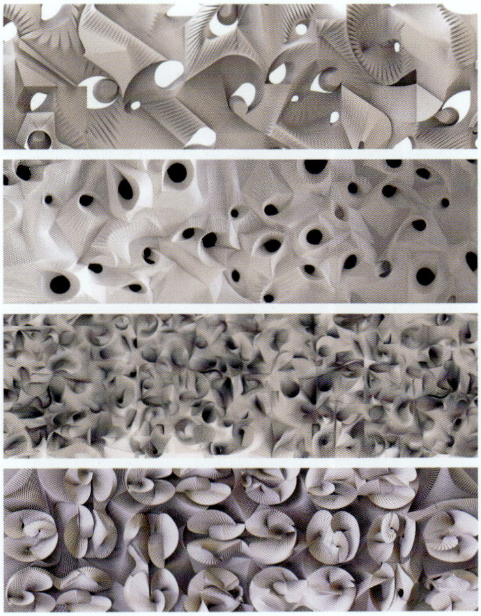

e

*models on plinths and fabricated ceiling prototypes mounted to the ceiling. Students: a) Xiangguo Cui, Siyu Dong, Davis Dunaway, Jingjing Yan, b) Adhityan Anbumozhi, Cem Akgun, Pouria Vakhshouri, Yinglei (Amber) Chen, Sihan Li. c-e) Rows (1) Grey Wartinger, Jiansong (Ricky) Yuan, (2) Christine Khouri Sader, Vicny Yang, Ryan Liao, (3) Deon Kim, Riley STudebaker, and (4) Geng Liu, Yuran Liu.*

must develop these electrical circuits and lighting elements in such as way as to heighten the aesthetic properties of their robotically fabricated ceiling prototypes.

In this course, students build upon a long tradition of architectural lighting, using their designs to heighten forms of attention, accentuate formal qualities, and engage with viewers in novel ways. Using direct or indirect lighting, their interactive lighting systems aim to draw viewers in and modulate their behavior or add expressive qualities to their ceiling prototypes. Ultimately, students use these projects to speculate on autonomous and persistent behaviors in an architecture that operates as a cyberphysical system, expanding design agency in architectural ceiling surfaces.

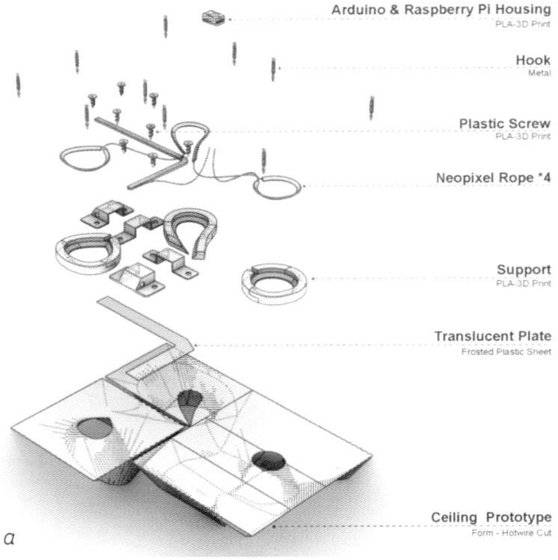

*a) Exploded view of cyberphysical systems components for ceiling, and b) Close-up of robotically fabricated ceiling prototype with wire-cut tooling articulation. Students: a) Jingyu (Joey) Luo, Shunta Moriuchi, Shuoxuan (Shane) Su, Renhu (Franklin) Wu, b) Deon Kim, and Riley STudebaker.*

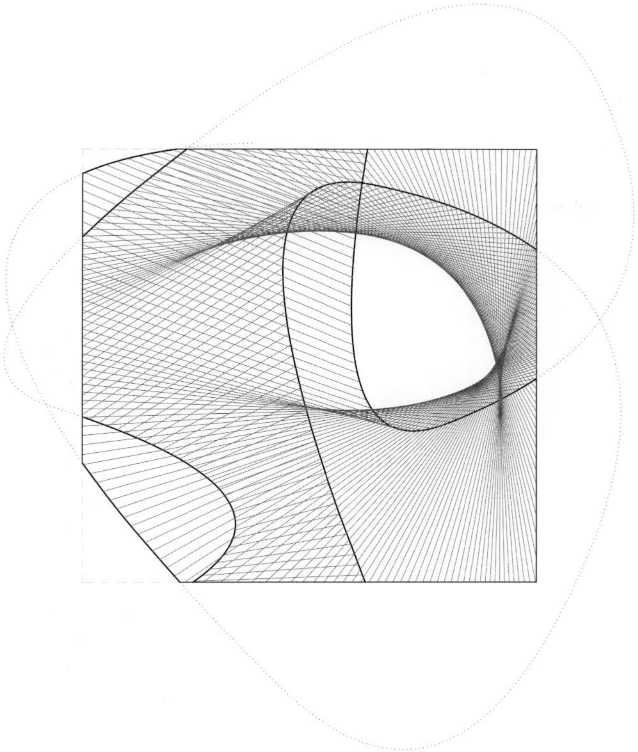

*a*

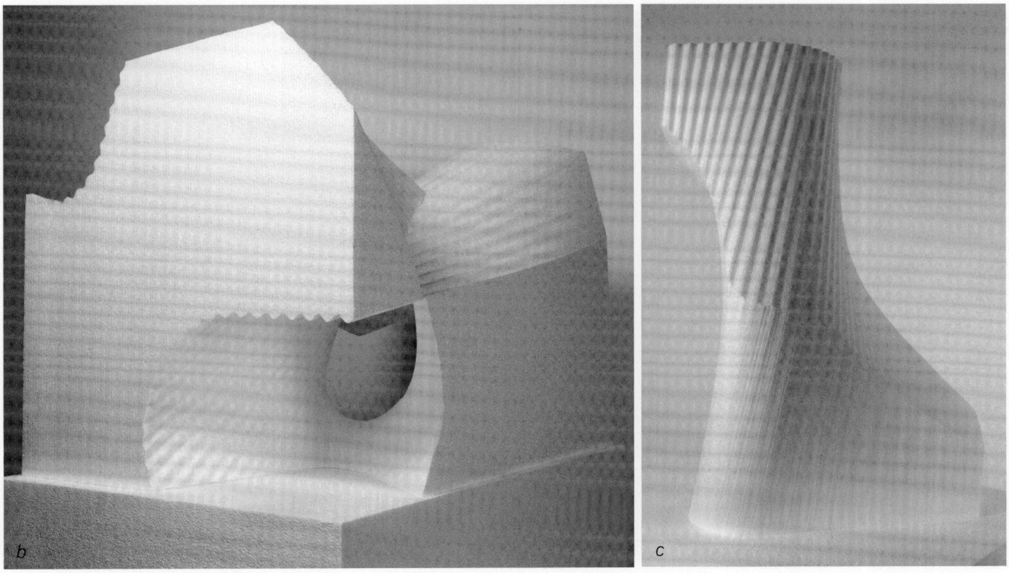

*a) Ruled-surface wireframe and robot hot-wire cutting toolpath for a volumetric study of a Pevsner/Gabo sculpture, b,c) EPS positive and negative volumes resulting from robot hot-wire cut volumetric study. Students: Claire Moriarty, and Matthew White.*

# FALL SECTION I JURORS

2020-2022

***Guest Jurors:***

Viola Ago *Rice*
Dorit Aviv *Penn WSOD*
Joseph Choma *Clemson*
Karl Daubmann *LTU*
Stella Dourtme *SOM*
Winka Dubbeldam *Penn WSOD*
Scott Erdy *Penn WSOD*
Gabriel Esquivel *Texas A+M*
Jelle Feringa *Aectual*
Kyriaki Goti *Pratt*
Elena Mangigian *Erdy McHenry*
Frank Melendez *CCNY*
Sharvari Mhatre *Penn WSOD*
Drew Ketz *Erdy McHenry*
Hyojin Kwon *Harvard GSD*
M.Casey Rehm *Sci-Arc*
Laia Mogas Soldevila *Penn WSOD*
Frederik Steiner *Penn WSOD*
Robert Stuart-Smith *Penn WSOD*
Kathy Velnikov *Taubman*
Danielle Willems *Penn WSOD*

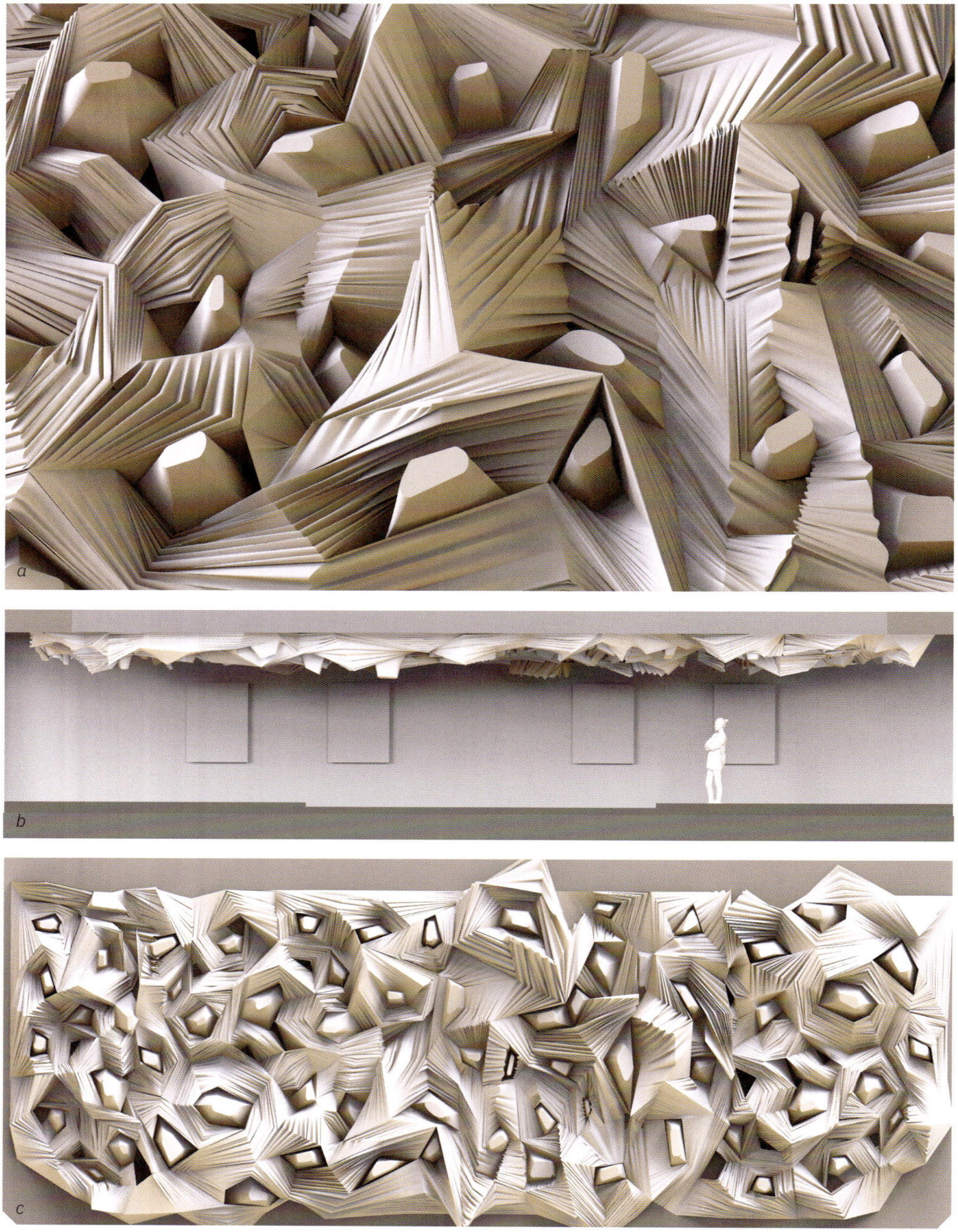


*Students: Xiangguo Cui, Siyu Dong, Davis Dunaway, and Jingjing Yan*

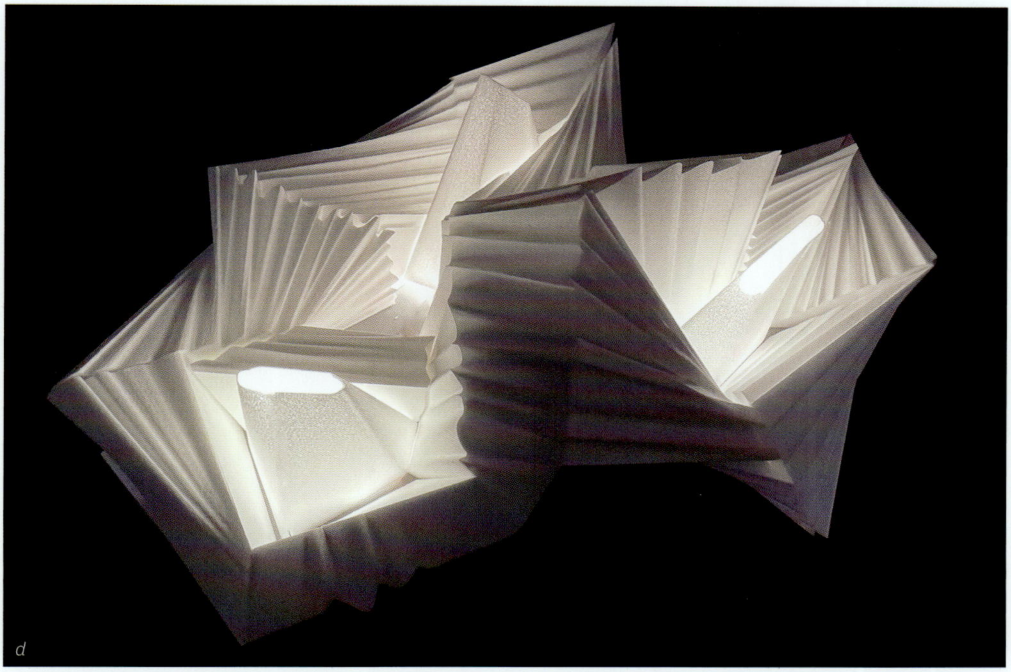

*a) Reflected ceiling plan close-up, b) long elevation of ceilingscape, c) reflected ceiling plan, and d,e) photographs of robotically fabricated prototype*

*Students: Grey Wartinger, and Jiansong (Ricky) Yuan*

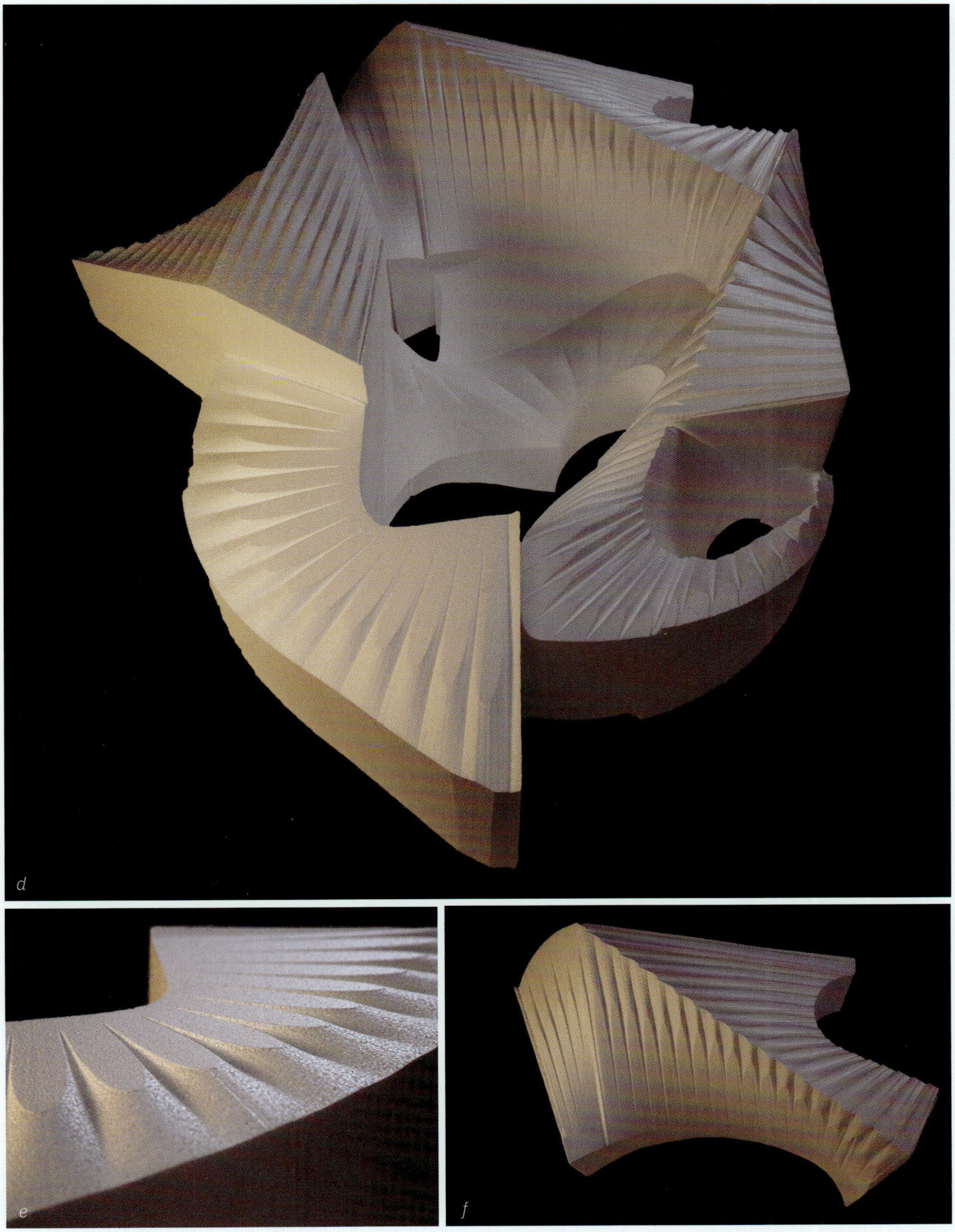

*a) Interior perspective of ceilingscape, b) long elevation of ceilingscape, c) reflected ceiling plan, and d,e,f) photographs of robotically fabricated prototype*

*Students: Christine Khouri Sader, Vicny Yang, and Ryan Liao*

d

e

*a) Interior perspective of ceilingscape, b) long elevation of ceilingscape, c) reflected ceiling plan, and d,e,f) photographs of robotically fabricated prototype*

a

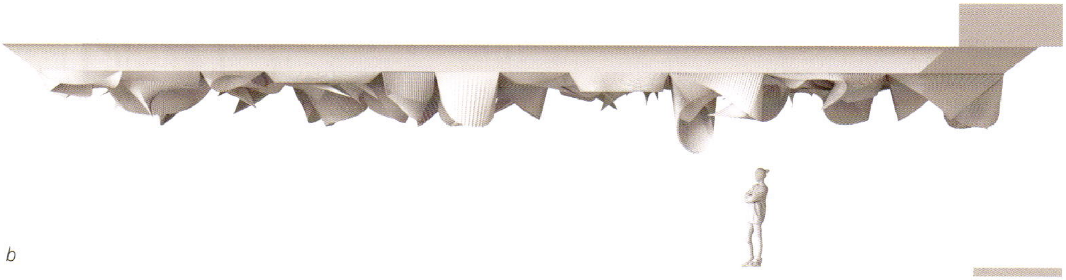

b

c

Students: Jingyu (Joey) Luo, Shunta Moriuchi, Shuoxuan (Shane) Su, and Renhu (Franklin) Wu

*a) Interior perspective of ceilingscape, b) long elevation of ceilingscape, c) reflected ceiling plan, and d,e,f) photographs of robotically fabricated prototype*

*Students: June Mingyang Yuan, Chenxiao Li, Yuxuan Wang, and Zilong Han*

*a,b) Photographs of robotically fabricated prototype, c) interior perspective of ceilingscape, and d) reflected ceiling plan.*

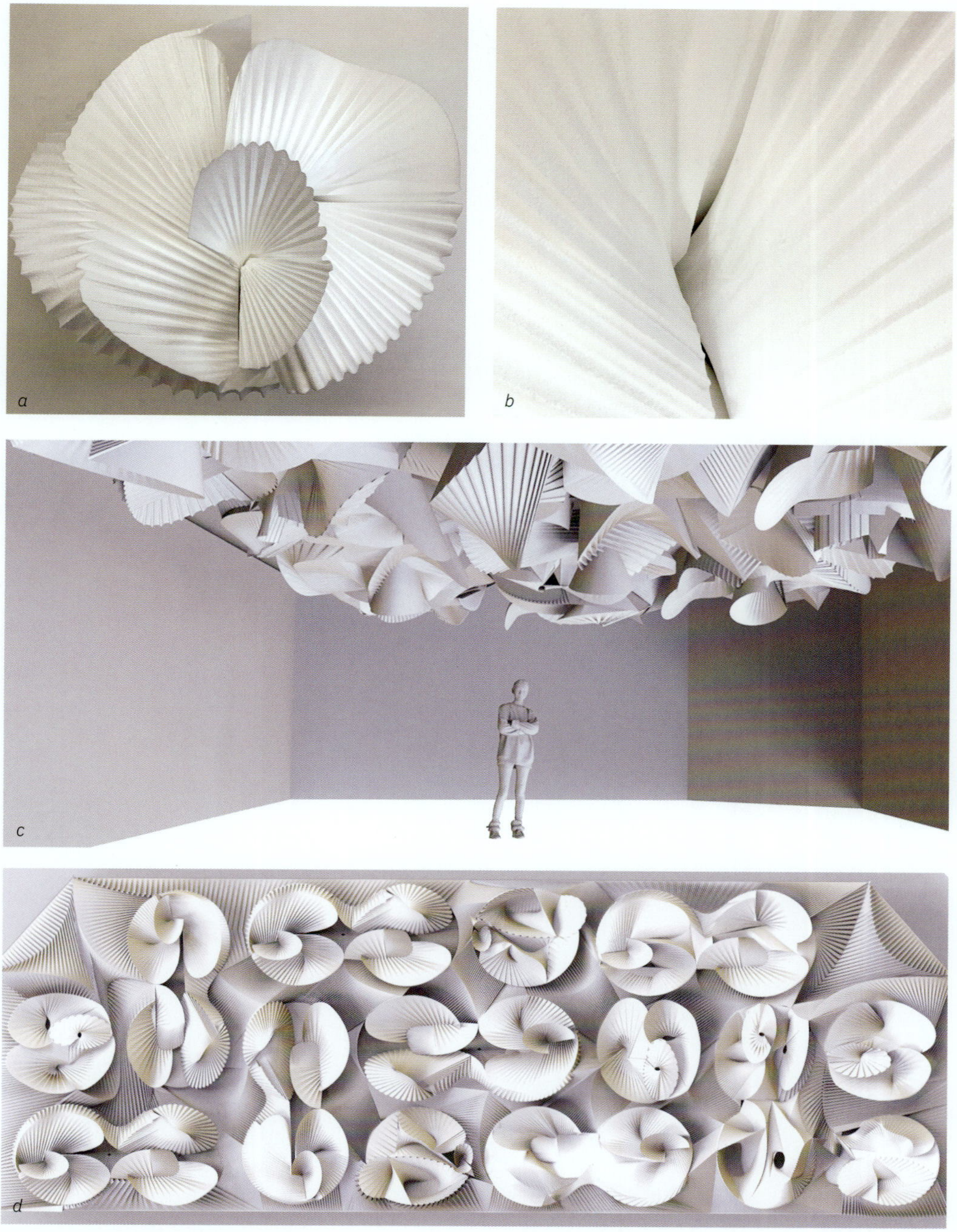

*Students: Geng Liu, and Yuran Liu*

*a,b) Photographs of robotically fabricated prototype, c) interior perspective of ceilingscape, and d) reflected ceiling plan.*

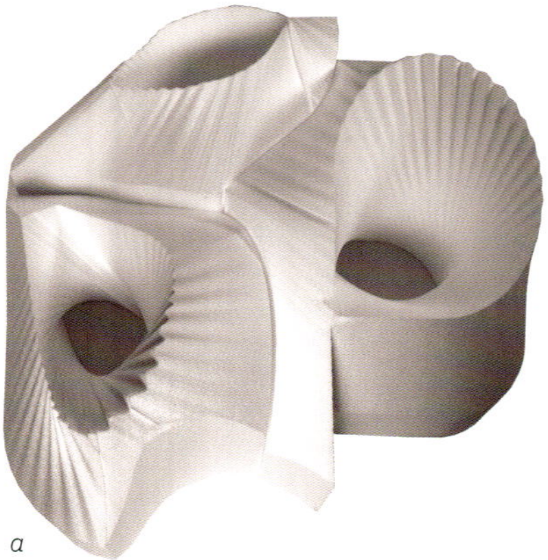

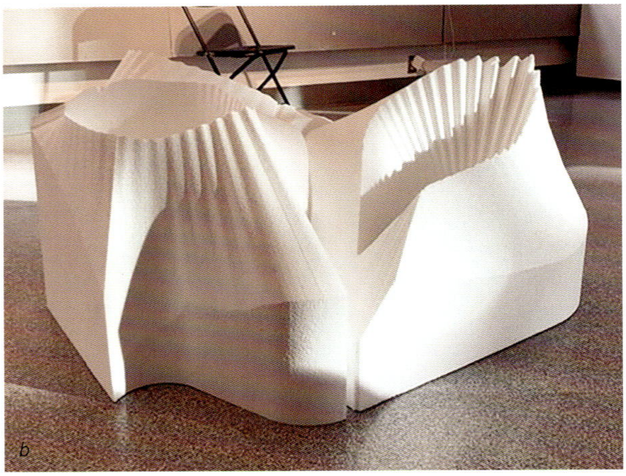

*Students: Claire Moriarty, and Matthew White*

*a,b) Photographs of robotically fabricated prototype, c) interior perspective of ceilingscape, and d) reflected ceiling plan.*

*Students: Layton Gwinn, Sophia O'neill, and Dan Rothbart*

*a,b) Photographs of robotically fabricated prototype, c-d) volumetric translation studies of Pevsner/Gabo sculptures, e) interior perspective of ceilingscape, and f) reflected ceiling plan.*

*Students:Deon Kim, and Riley Studebaker*

*a,b) Photographs of robotically fabricated prototype, c) interior elevation of ceilingscape, d-e) Convolutional Neural Network generated images, and f) reflected ceiling plan.*

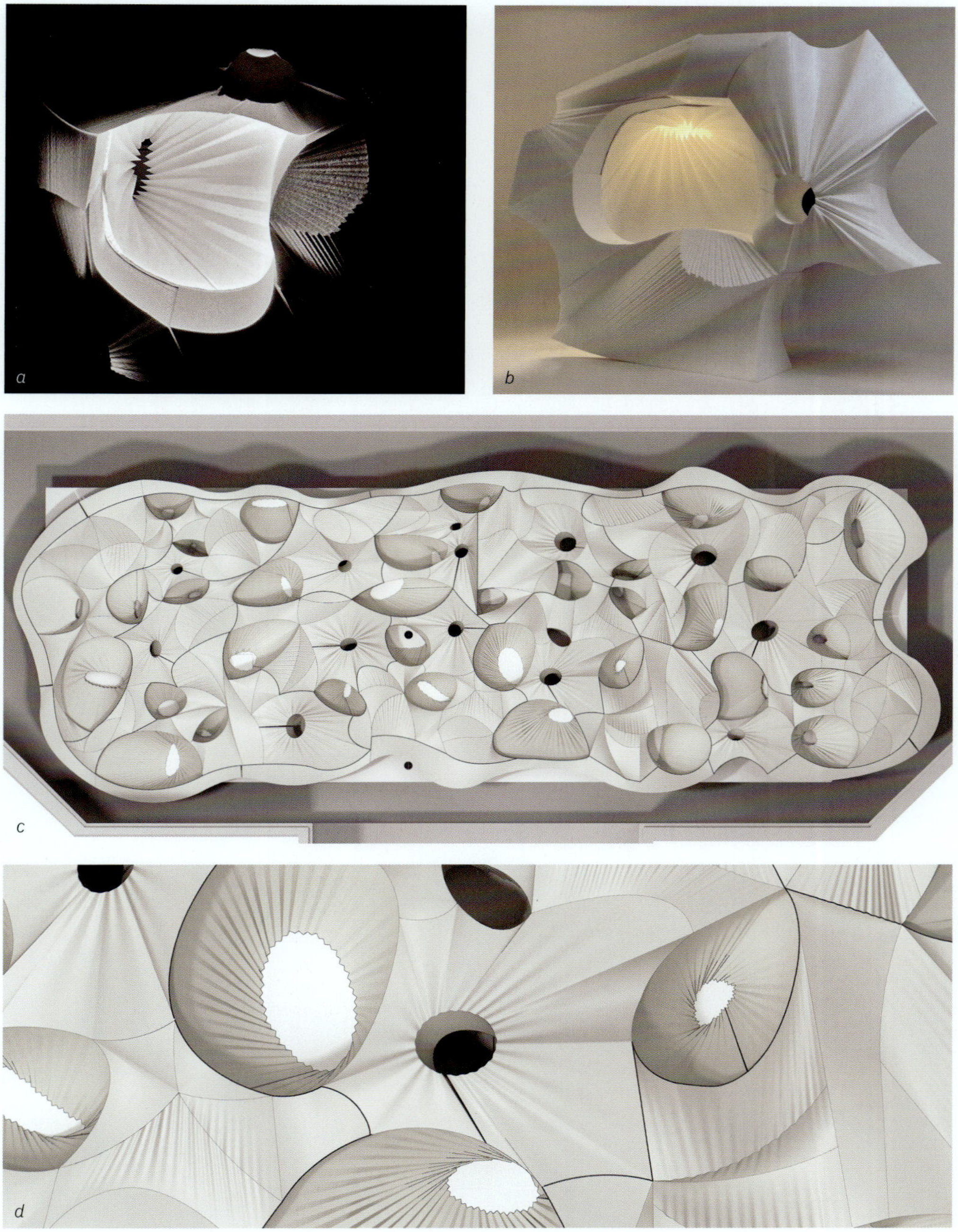

Students: Adam Blood, Aisha Alshehri, Hrishi Rajasekar, and Matt Ward

a,b) Photographs of robotically fabricated prototype, c) reflected ceiling plan, and d) close-up reflected ceiling plan.

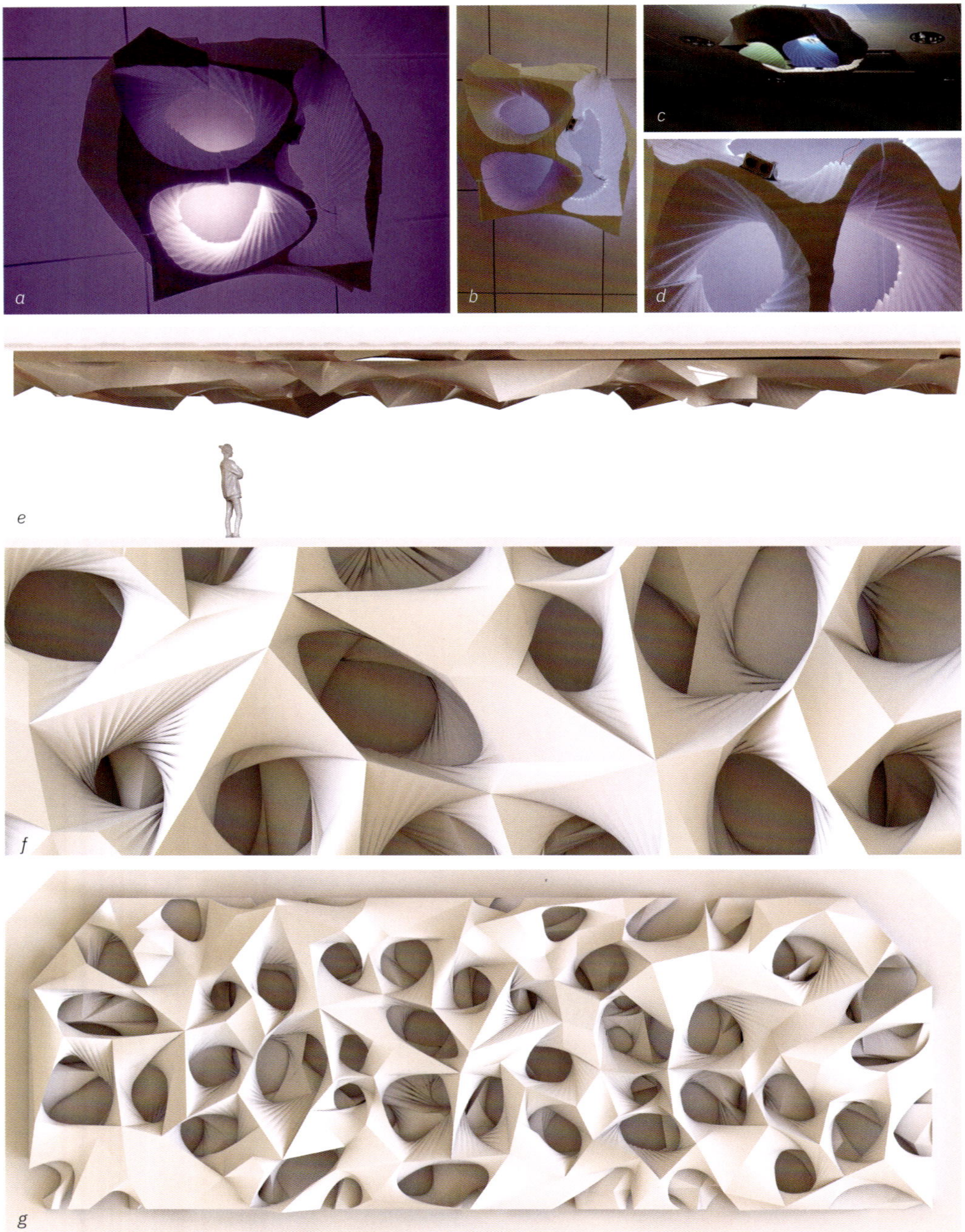

*Students: Adhityan Anbumozhi, Cem Akgun, Pouria Vakhshouri, Yinglei (Amber) Chen, and Sihan Li*

*a-d) Photographs of robotically fabricated prototype, e) Long elevation, f) reflected ceiling plan close-up, and g) reflected ceiling plan.*

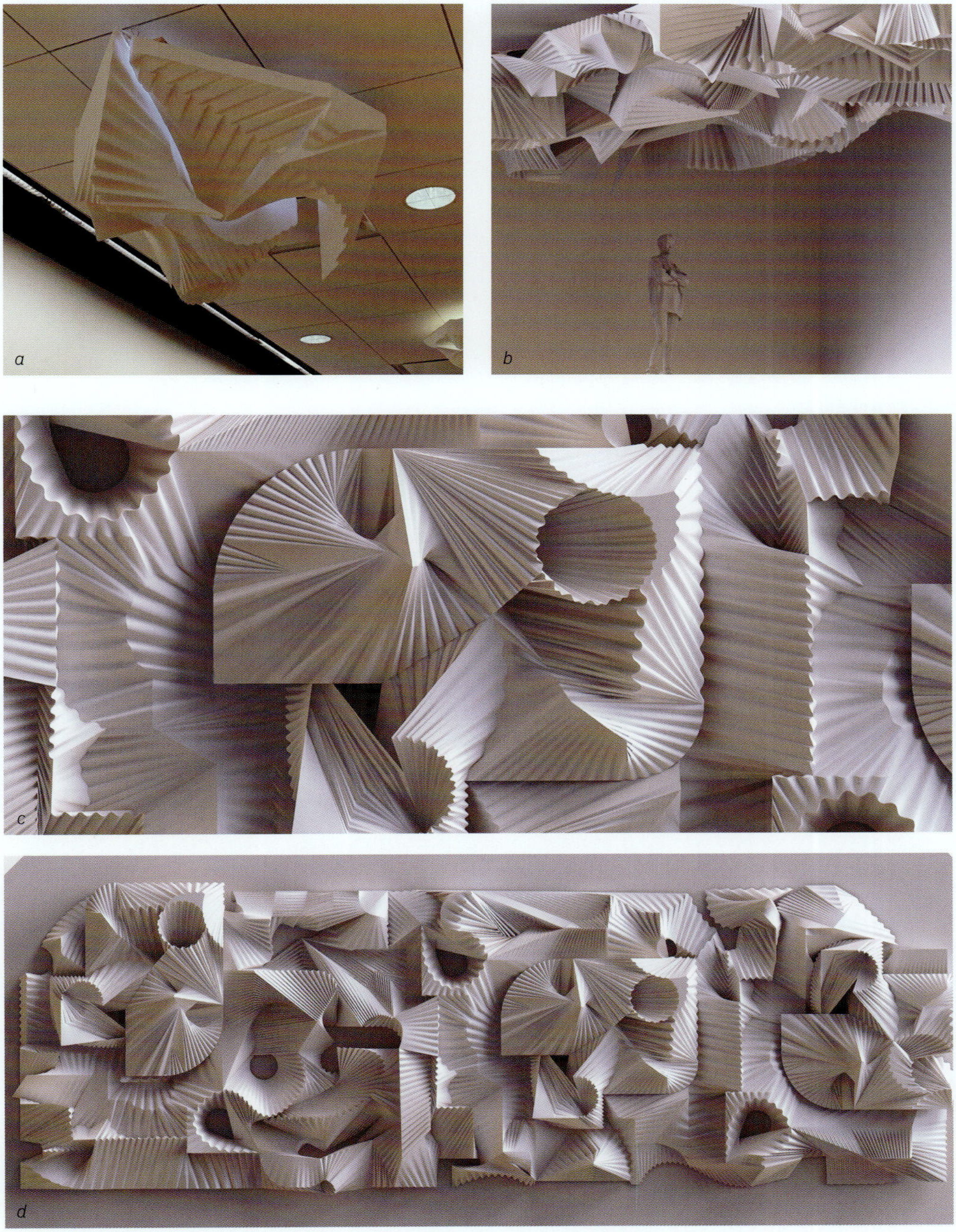

*Students: Chunze Li, Haohan Tang, Yidan Wang, and Bentian Wang*

*a,b) Photographs of robotically fabricated prototype, c) reflected ceiling plan close-up, and d) reflected ceiling plan.*

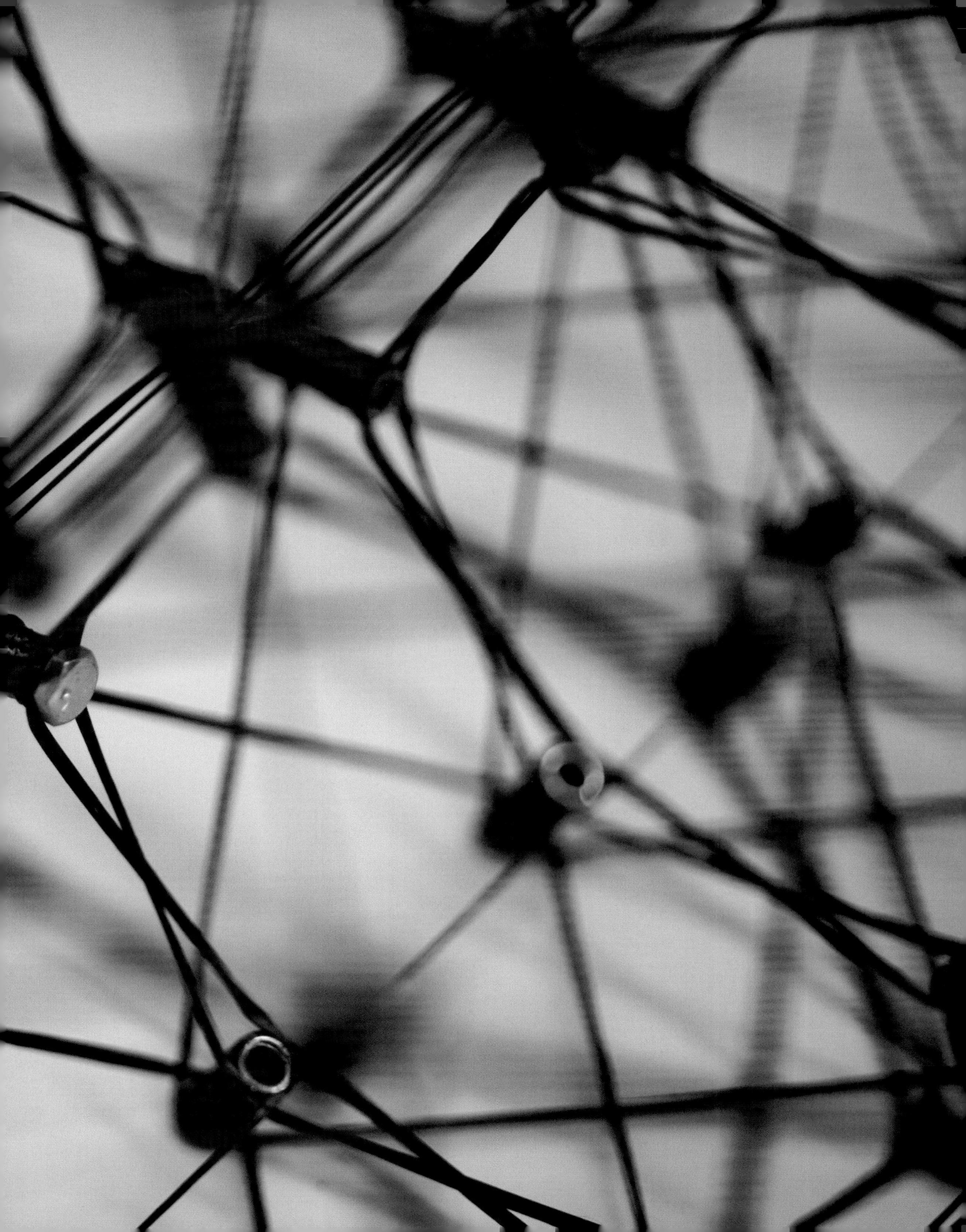

# FALL SECTION II

**The second half of the fall semester explored designing and prototyping a fiber-composite aggregate assemblage. Students developed robot tool paths and jigs for fiber composite filament winding and generative algorithmic design approaches to spatial aggregation and the organization of filament-wound fiber in designed volumetric aggregates. This work was undertaken within both a design studio and a technology course.**

## 8011 Material Agencies: Robotics and Design Lab II

Instructor: Ezio Blasetti
TAs: Patrick Danahy, Geng Liu, Yuan Mingyang, and Sophia O'Neil

Material Agencies focuses on the design and robotic fabrication of a fiber-composite prototype developed through generative algorithmic methods of architectural design. Fiber composite parts were manufactured by robotic filament-winding a carbon-fiber thread around a custom-designed jig. The carbon-fiber thread is pre-impregnated with a resin. Post-winding threaded parts must be baked in an oven to cure the resin into a hard material that, combined with the carbon fiber, forms a high-strength and lightweight composite. Given the volumetric constraints of an industrial oven, projects explored the design and manufacturing of an aggregate-based proposal, where an individual building block or its constituent parts are scaled to fit into the oven.

*a) Robotic filament-wound prototypical assemblage, and b-c) robotic filament-winding over a reconfigurable jig. Students: Xiangguo Cui, Siyu Dong, Davis Dunaway, Jingjing Yan | June Yuan, Chenxiao Li, Yuxuan Wang, and Zilong Han.*

***Building block:*** Working in small teams, students designed three-dimensional aggregates and a custom jig with pegs, which could be used to fabricate the aggregates through robotic filament winding. Experiments were conducted to develop appropriate robot winding programs and the jig to ensure designs adhered to robot kinematic constraints when maneuvering around the jigs. The fiber-placement within each building block and their corresponding robot tool paths, together with a larger assemblage of several building blocks, was produced using custom algorithmic methods developed in the 803 course running parallel to Material Agencies.

***Design Brief:*** Participants developed speculative design proposals for an "Autonomous Public Garden." Historically, the space of the Garden is a hybrid, a gradient of public and private, a territorial buffer that interfaces with the commons and the environment. The Garden is an archetypal diagram of a network open to ecological, cultural, and economic feedback mechanisms, capable of adapting to change over time. Speculative design projects imagined ways to embody a contradictory nature: being at once always complete, yet always under construction. Operating as a lightweight endoskeleton, proposals sought to provide a scaffold/envelope that might support the growth of an organic and ephemeral system. This synthetic environment is envisaged to experience growth over time from both biological matter and adaptations to the tectonic aggregate system itself, with the two converging towards a single ecology. By remaining in a dynamic, unfinished state, each design's scale and organization of parts challenges architecture's traditional part-to-whole relationships, with distributed fibrous material orders producing more ambiguous relations between incomplete architectural wholes and their constituent aggregates. In this manner, projects speculated on the limits of architecture to operate as an organizational, structural, and aesthetic agent of change.

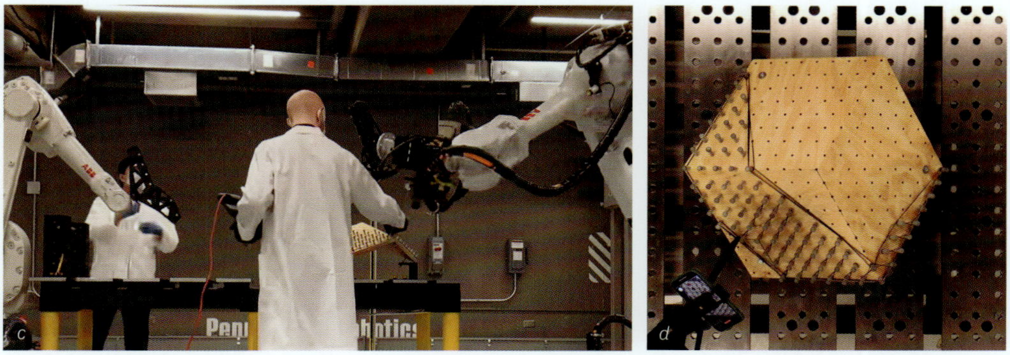


*a) Robotic filament-winding, and b) custom-developed 3D jig for filament winding. Students: Deon Kim and Riley Studebaker.*

## Introduction to Algorithmic Design & Robotic Fabrication

Instructor: Ezio Blasetti
TAs: Geng Liu, and Sophia O'Neil

A generative computational approach to the design of a fiber-composite volumetric part and the organization of several parts into a larger aggregate was undertaken to support the 8011 Material Agencies design brief. Teams designed a volumetric part geometry and fiber placement within this part through the development of a computational design methodology that sought to produce lightweight, structurally performative parts whilst integrating constraints and opportunities from the robot filament-winding manufacturing approach. Filament-winding methods involved both the design and fabrication of a jig with an array of pegs and an algorithm that determined a sequence of winding operations around these pegs to wrap the filament. This winding sequence operated as a robot tool path, programming an industrial six-axes robot arm to manufacture the part. Following robotic production of the pre-impregnated thread, the part is cured in an oven for several hours before being cooled and removed from the jig. Algorithms such as an agent or turtle-based motion were developed to simulate the winding sequence, producing a continuous path that built up a varied density and orientation of fibers in different regions of the volumetric part.

Explorations into the aggregation of several parts into a larger architectural assemblage questioned the spatial and structural organization of parts to form a viable whole and the reconfiguration and addition of parts over time. Algorithms that explored the distribution, orientation and growth of an aggregate system were explored through both indexical and non-indexical means. Examples include cellular automata, recursive aggregation, agent-based and genetic algorithms, often coupled with structural and/or environmental analysis in the loop. These algorithms were utilized to develop schematic design proposals although they could also be adapted to operate within a speculative built scenario where the computational method would enable proposals to operate as a programmable structure, a networked spatial computer that could be reconfigured to varied conditions over time.

a

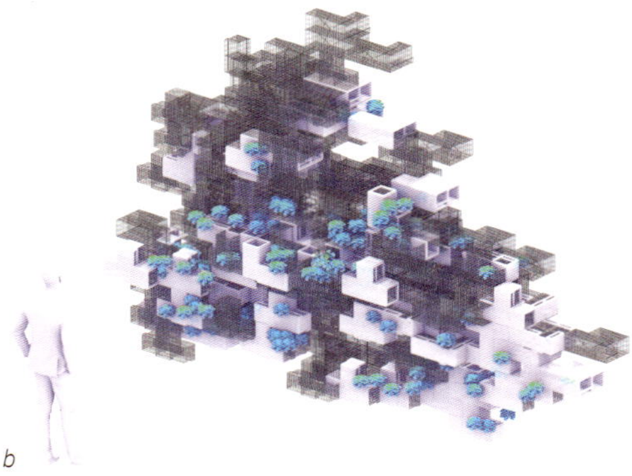

b

*Computational approaches to design and material placement. a) A polyscalar space-filling generative aggregation in relation to a topological structural optimisation, b) a Cellular Automata growth algorithm that distributes both solid and porous building blocks, and c) algorithmic material placement studies for robotic carbon-fiber filamnet winding. Students: a) Deon Kim, Riley Studebaker, b) Geng Liu, Yuran Liu, and c) Matthew White, Claire Moriarty, Zilong Han*

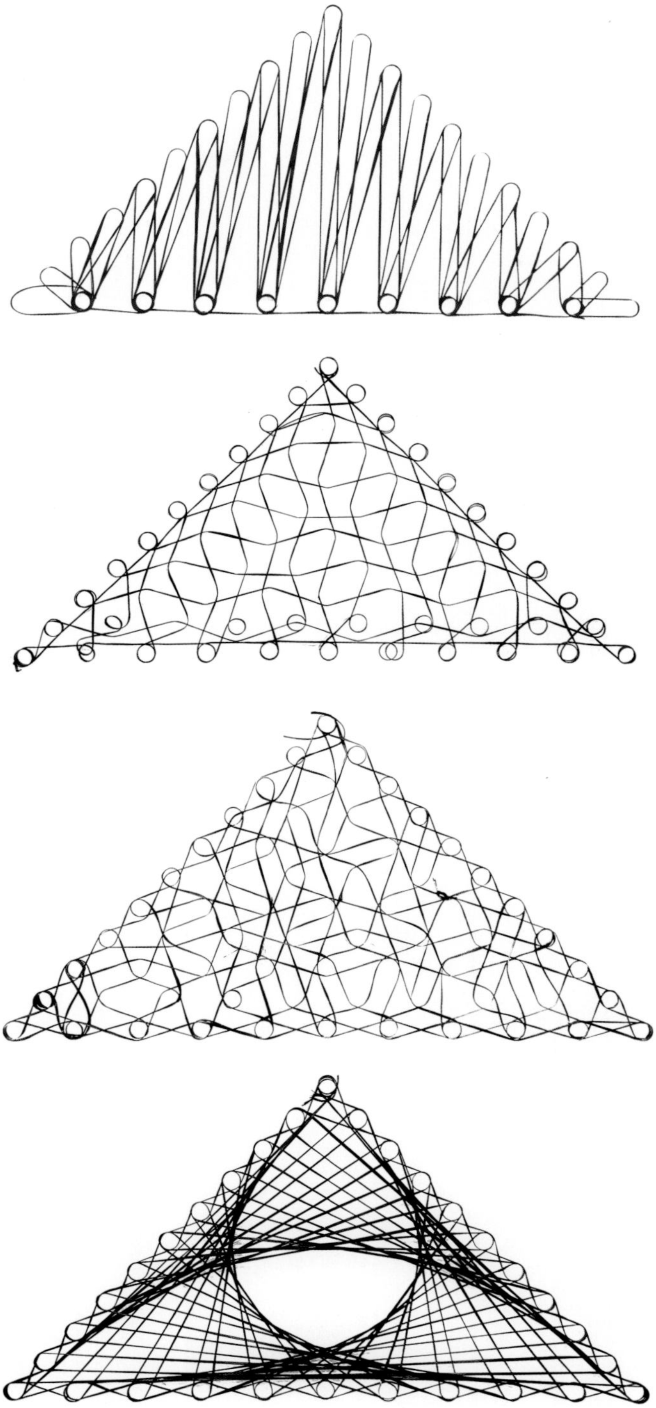

c

# FALL SECTION II JURORS

2020-2022

***Guest Jurors:***

Dorit Aviv *Penn WSOD*
Masoud Akbarzadeh *Penn WSOD*
Mohammad Bolhassani *CCNY*
Jonas Coersmeier *Penn WSOD*
Karl Chu *Metaxy*
Winka Dubbeldam *Penn WSOD*
Richard Garber *Penn WSOD*
Ferda Kolatan *Penn WSOD*
Stephan Henrich *Robotikdesgn/Arkitektur*
Ed Keller *AUM Studio*
Gokhan Kodalak *Cornell/Pratt*
Geoff Lemond *Lemond.com*
Greg Lemond *Lemond.com*
Sandra Manninger *SPAN/Taubman*
Timothy McCarthy *Lemond.com*
Philippe Morel *EZTC/ENSA*
Jose Sanchez *Plethora Project/Taubman*
Greg Sheward *Pratt Institute*
Robert Stuart-Smith *Penn WSOD*
Mathieu Victor *Via Domani*
Danielle Willems *Penn WSOD*

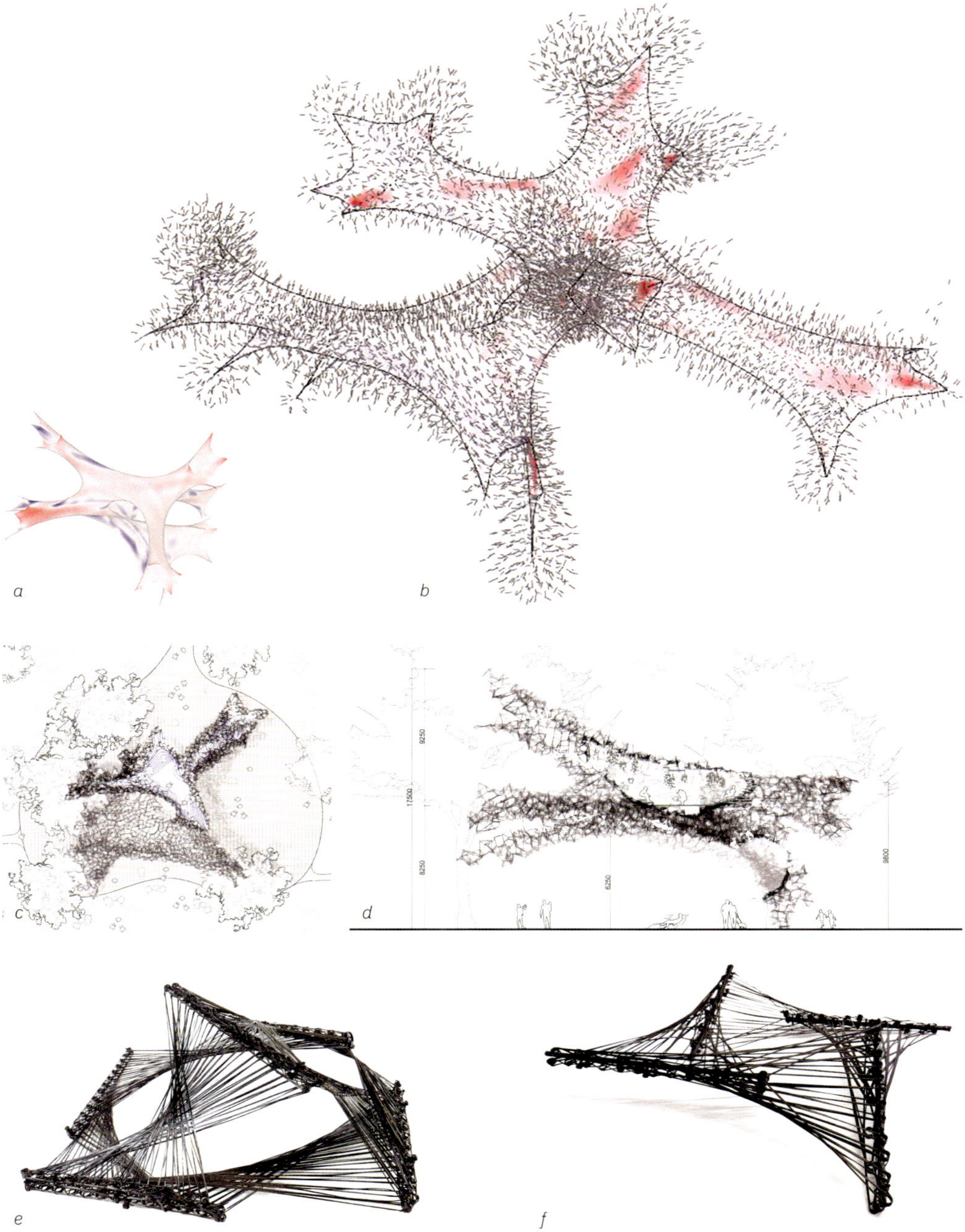


*Students: Xiangguo Cui, Siyu Dong, Davis Dunaway, and Jingjing Yan*

*A floating garden with public space is suspended between a cluster of trees. a) Structural Deflection Analysis, b) multi-agent material placement in relation to structural analysis, c) roof plan, d) elevation, e,f) initial prototypes of individual components, and g) multi-agent material-placement computer simulation*

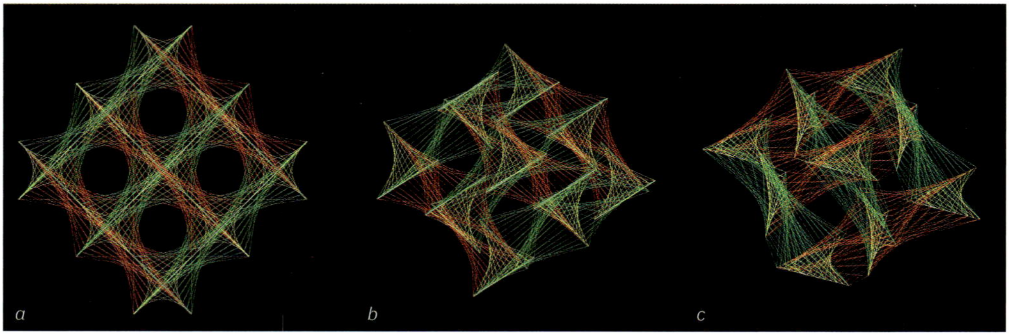

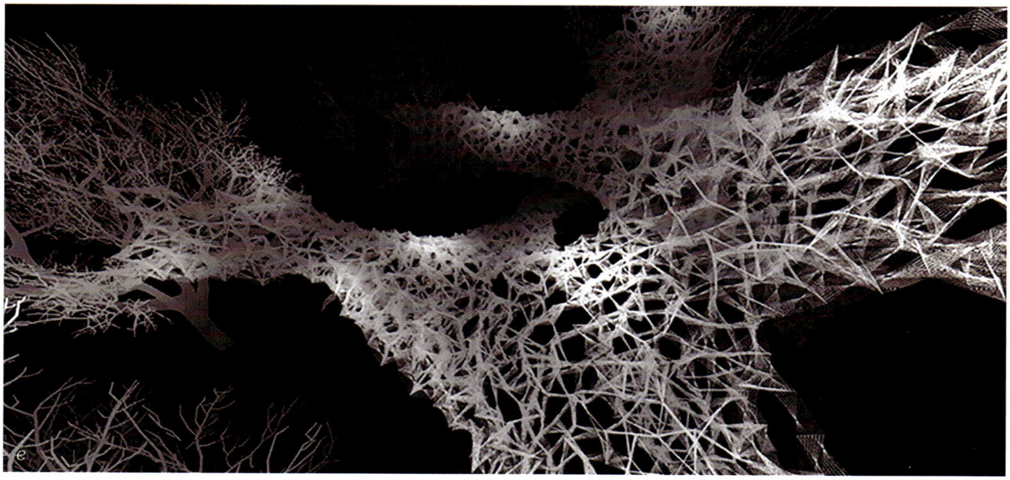

*Students: Xiangguo Cui, Siyu Dong, Davis Dunaway, and Jingjing Yan*

a,b,c) Filament winding organisation, d,f) exterior perspectives, e) perspective view of multi-agent material placement simulation, and g) photo of robotically fabricated final prototype.

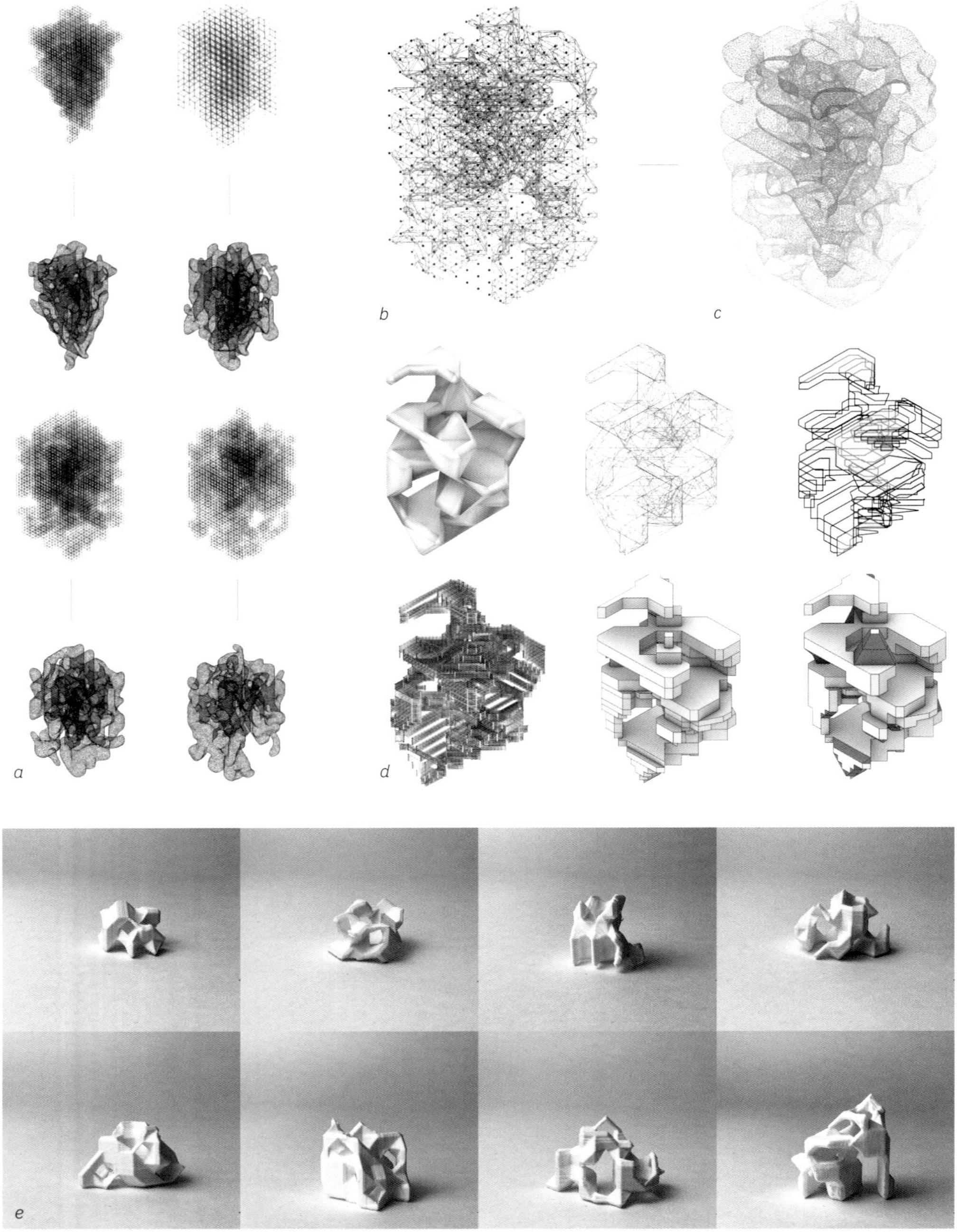


*Students: Jeffery Liao, Masalegoo Mahsa, Nicholas Houser, Xinlin Lu, and Tinghui Mo*

*f*

*A Cellular Automata Garden Grotto. a-c) A structurally stable Cellular Automata design process generates a point cloud organization that is interpolated as an isosurface, d) abstraction of the isosurface into a series of filament-wound stepped contoured trellis components, e) scale model 3D prints of different configurations of the grotto over time, and f) robotically fabricated prototype.*

a

b

*Students: Jeffery Liao, Masalegoo Mahsa, Nicholas Houser, Xinlin Lu, and Tinghui Mo*

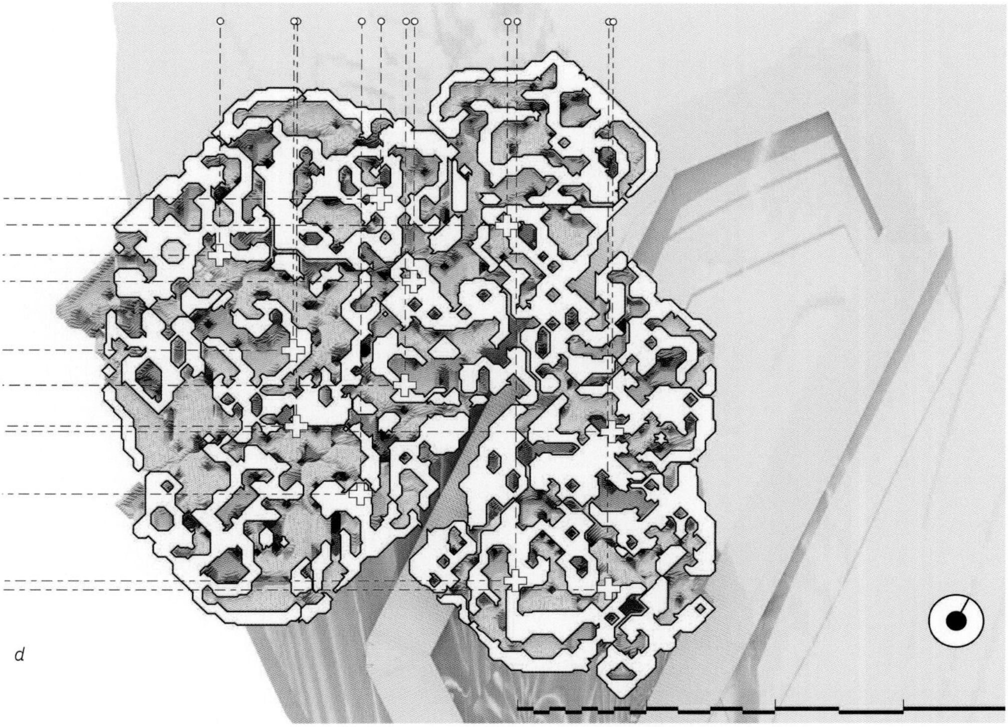

d

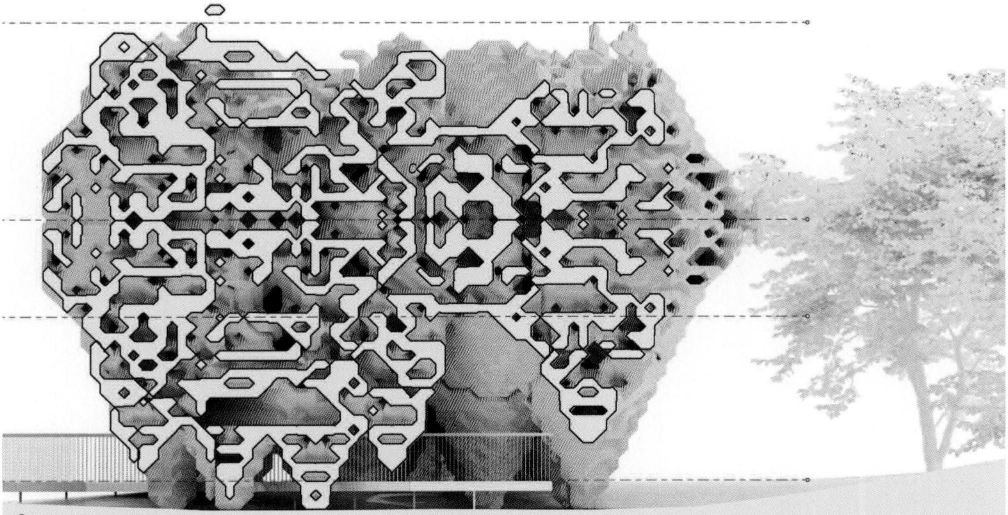

e

*a) Exploded perspective, b) exploded perspective of filament-wound prototype, c) exterior perspective, d) prototypical plan, and e) prototypical section illustrating the porous grotto spatial organization that aims to support rewilding plant-growth over time.*

Students: Matthew White, Claire Moriarty, and Zilong Han
A folded plate structure operates as both a green wall and rock-climbing wall. a) Computational investigation into folded plate structures, b) filament-wound carbon fiber studies of a single plate, c) several plates unfolded, d) a 3D set of folded plates, e-g) paper origami studies, h) close-up of folded filament wound parts ontop of a 3D mold, and i) perspective render of rock-climbing wall.

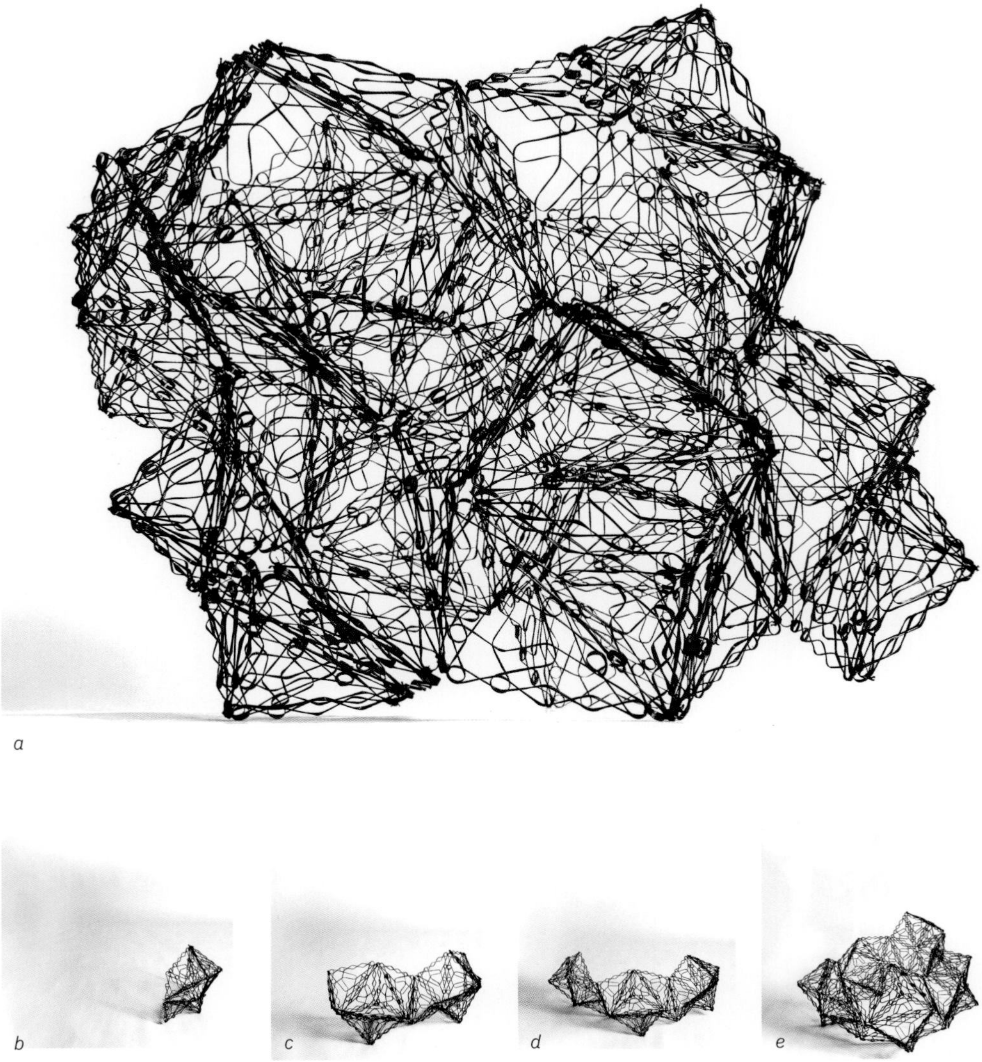

*a) Close-up of final volumetric assembly of planar filament-wound prepreg carbon-fiber, and b-e) progressive assembly of filament-wound carbon-fiber planes.*

*a*

*Students: Matthew White, Claire Moriarty, and Zilong Han*

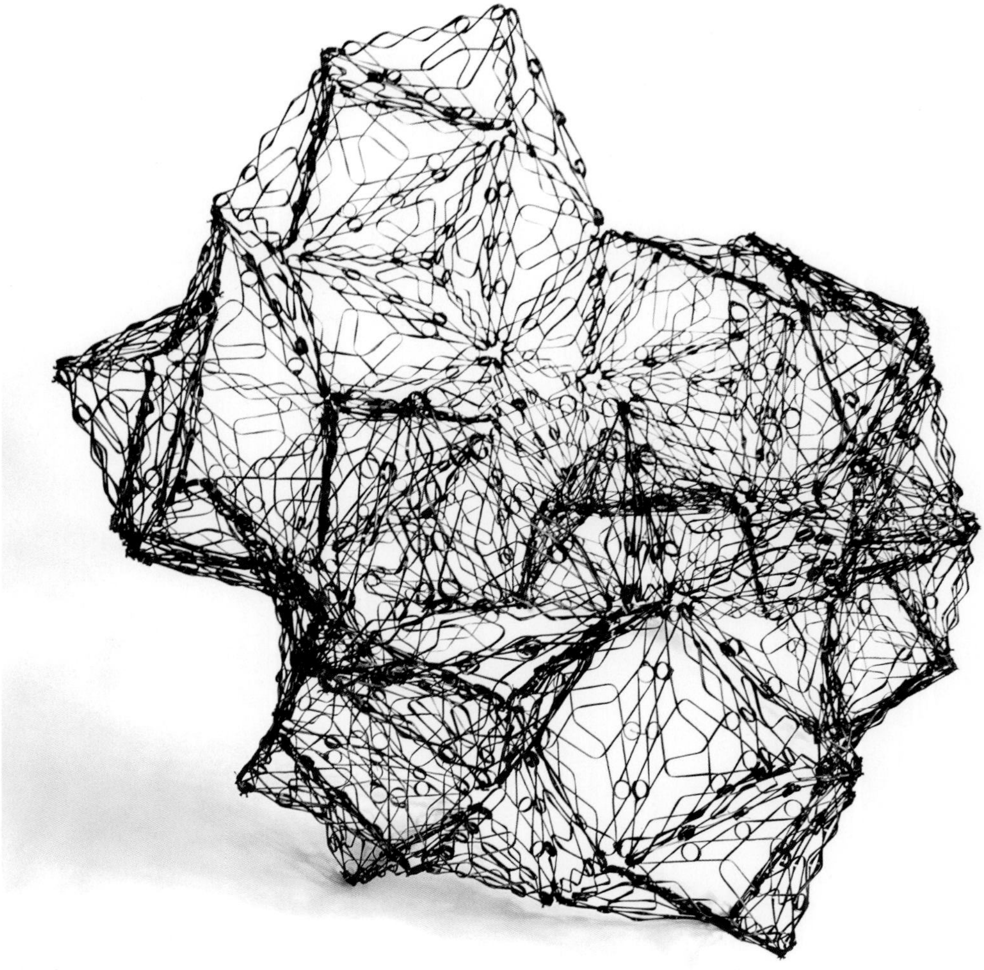

*b*

*a) 2D assembly of filament-wound prepreg carbon-fiber elements, and b) folded, 3D assembled multi-faceted planar filament-wound prototype.*

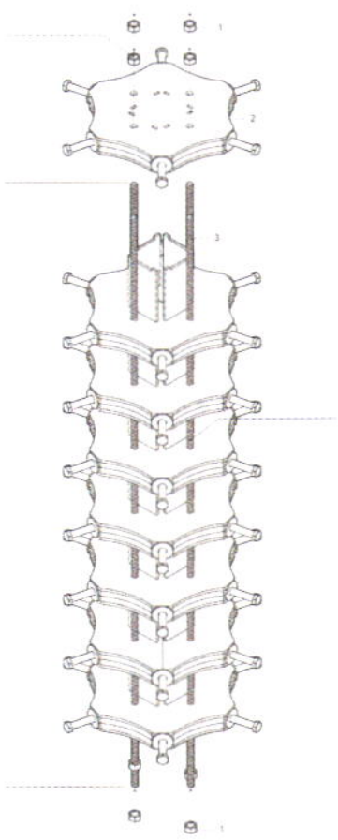

*a*

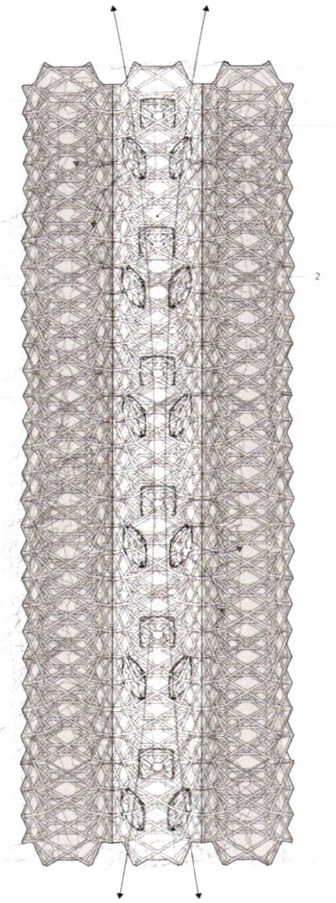

*b*

*c*

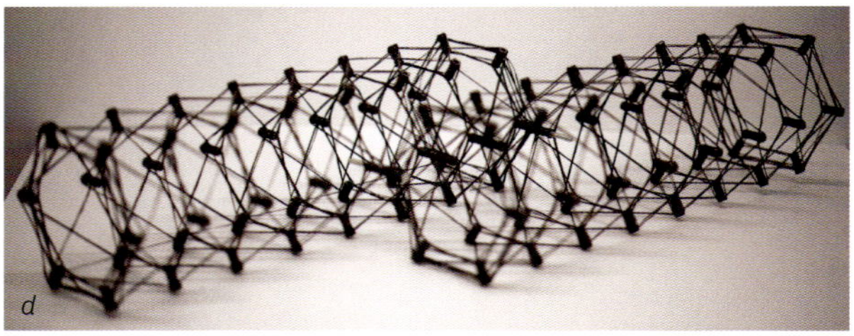

*d*

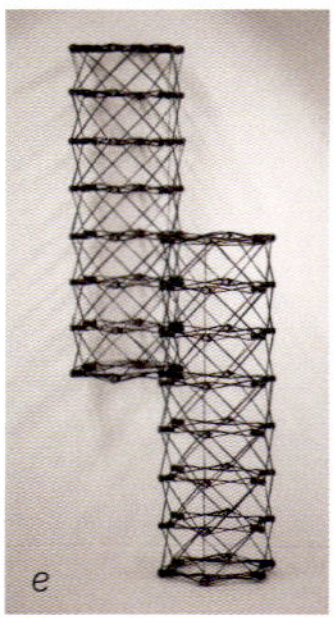

*e*

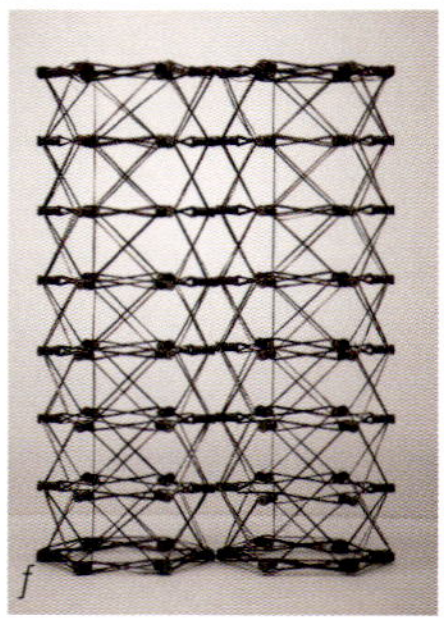

*f*

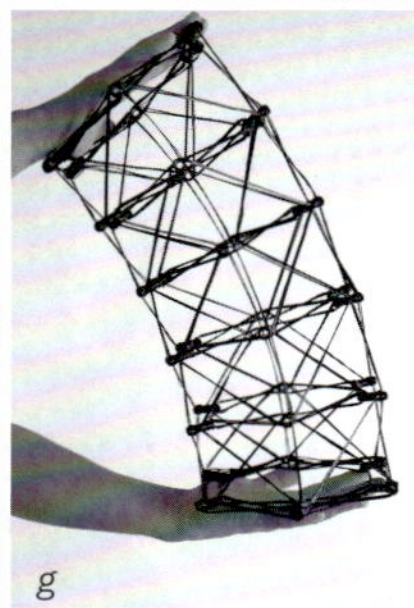

*g*

*Students: Grey Wartinger, and Jiansong Yuan*

*An ultralight filament-wound prepreg carbon fiber tubular-like elment is fabricated on a modular, reconfigurable, bendable jig. The method is envisaged to support the rhizomatic growth of a local garden that could be setup to locally fabricate and reconfigure these elements over time.*

*BUILDING*: FALL SECTION II

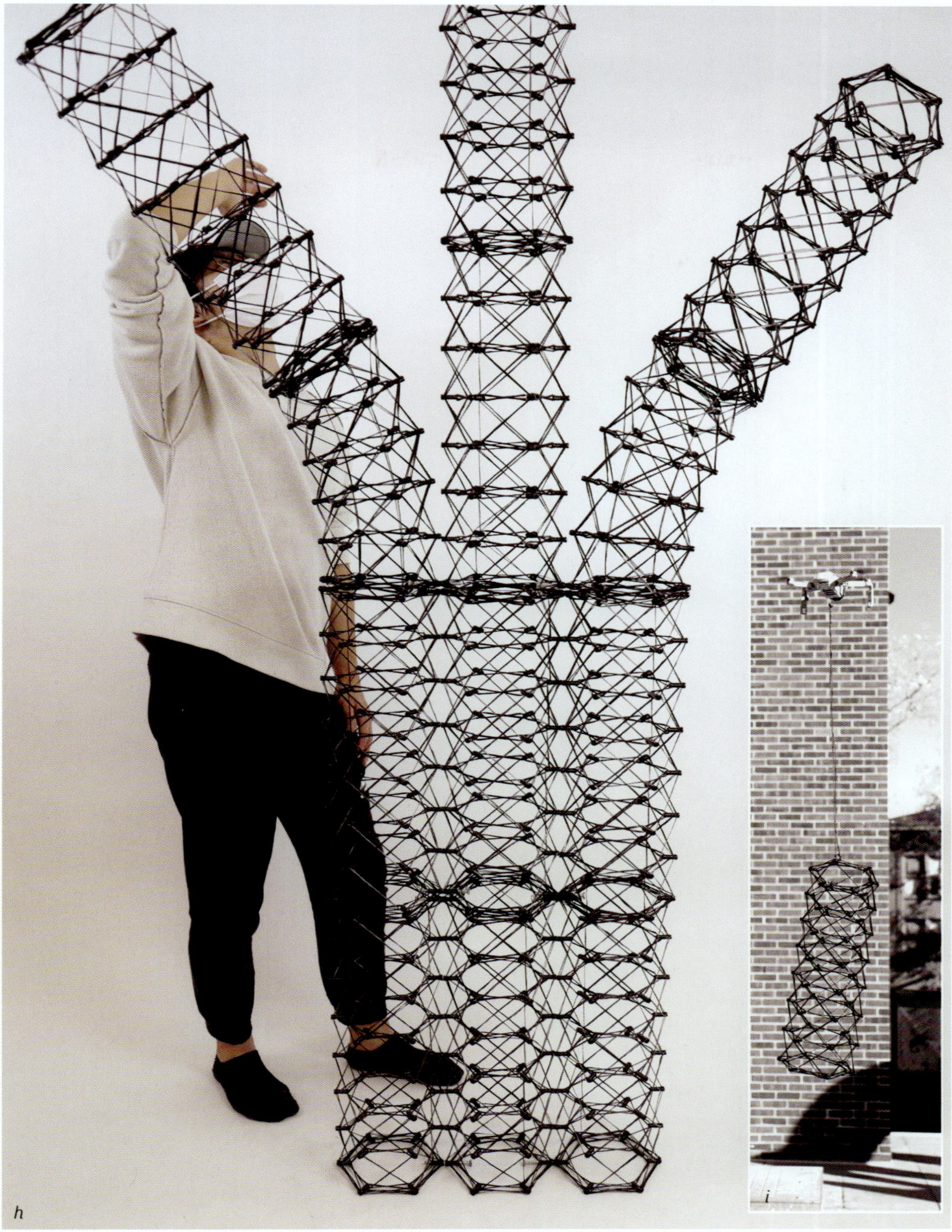

a) Filament-winding jig comprised of reconfigurable components, b) drawing of multi-component assemblage incorporating environmental sensors, c) hempcrete infil creates a hybrid material system similar to reinforced concrete, d-f) different configurations of filament-wound components, g) bent filament-wound component using jig, h) large multi-part assembled prototype comprised of robotic filament-wound components, and i) single component being air-lifted by a small recreational quadcopter.

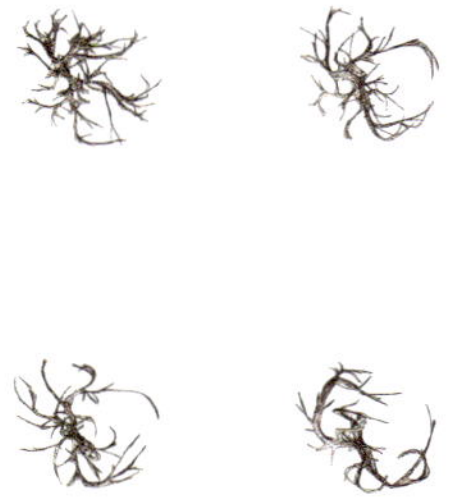

*a*

*b*

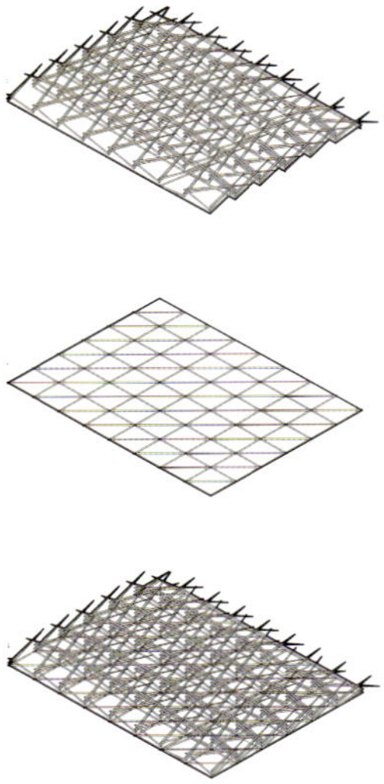

*c*

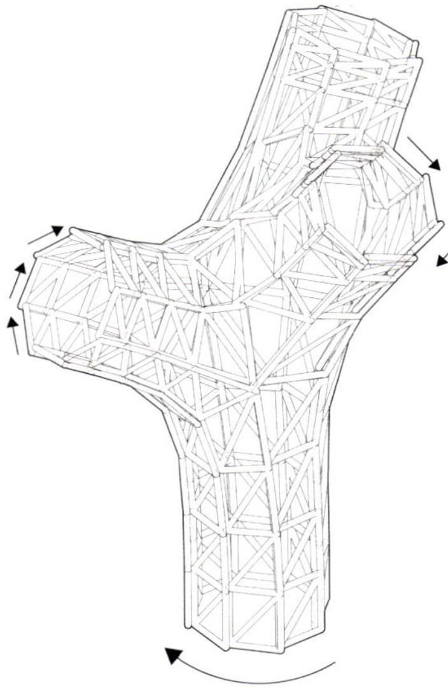

*e*

*Students: Haohan (Jason) Tang, Bentian Wang, Pouria Vakhshouri, and Hrishi Rajasekar*

*A porous, and tubular sponge-like garden is envisaged to enable plants to grow on its substrate, while providing generous space for light and air which also enables public space to weave around this 3D trellis garden installation.*

*BUILDING*: FALL SECTION II

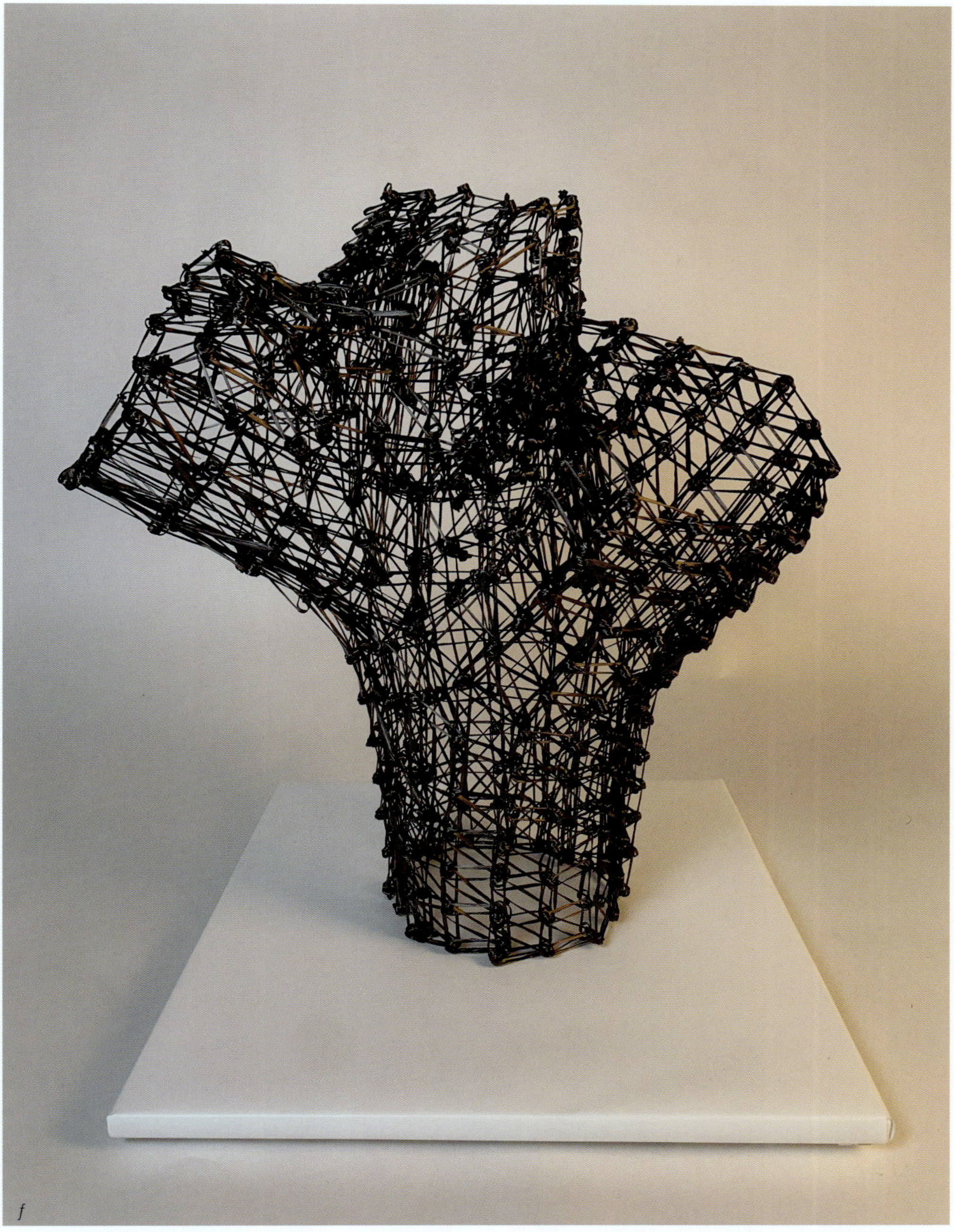

a) 3D massing growth algorithm, b) section through garden, c) filament-winding patterns and their translation on to portions of a branching tubular surface, d) rendered Perspective from inside the garden, e) porous tubular lattice supports plant growth, and f) robotic filament-wound prepreg carbon fiber prototype.

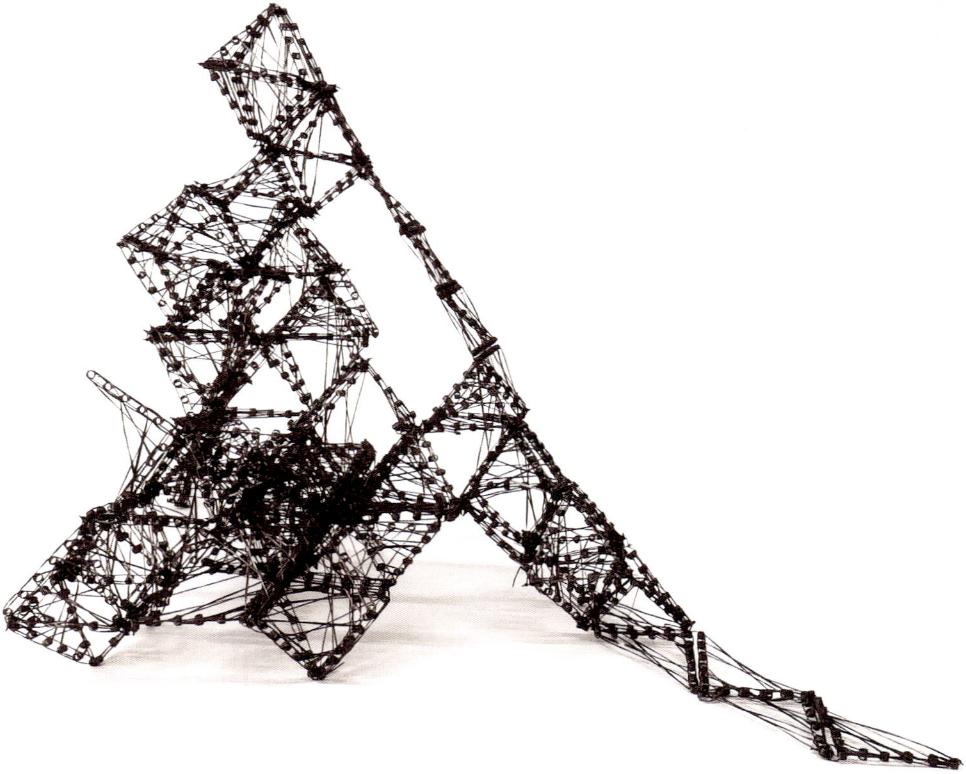

*a*

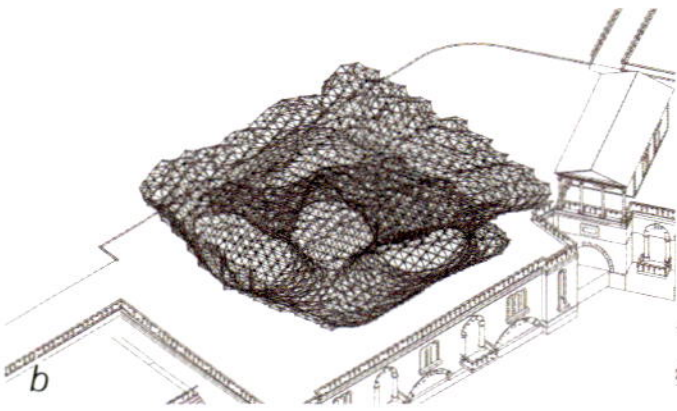

*b*

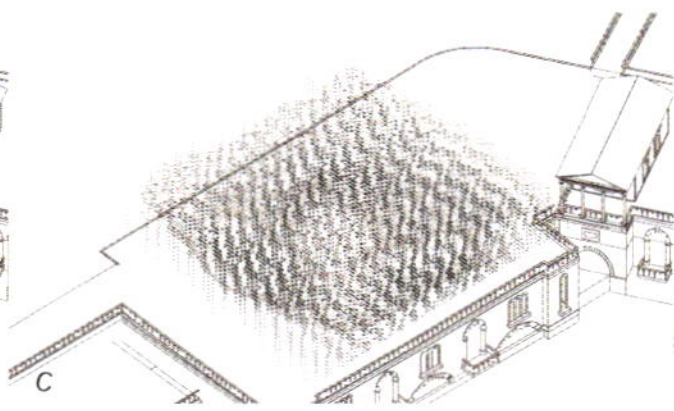

*c*

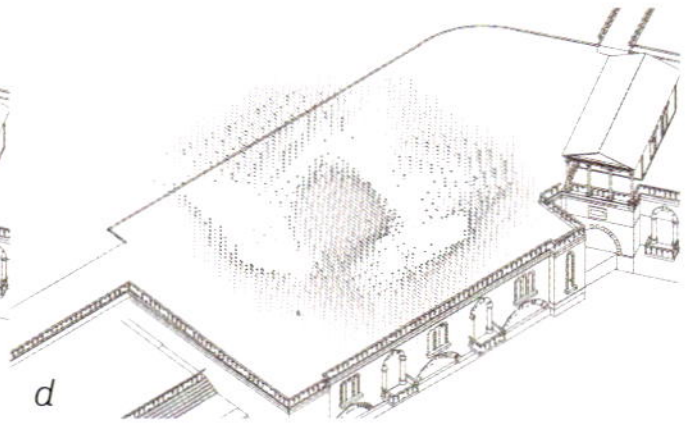

*d*

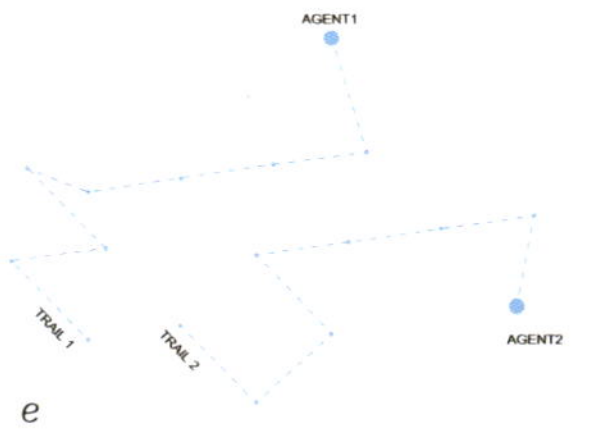


*e*

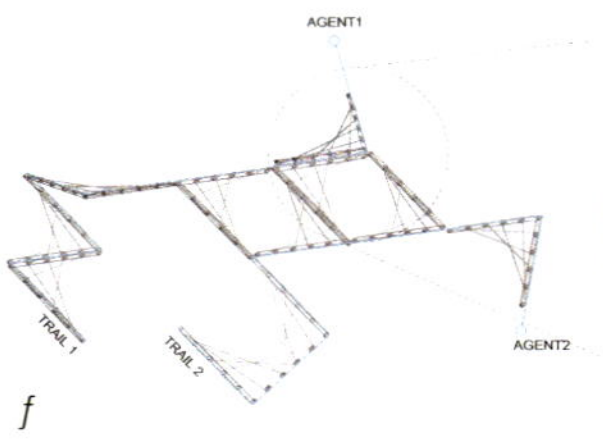


*f*

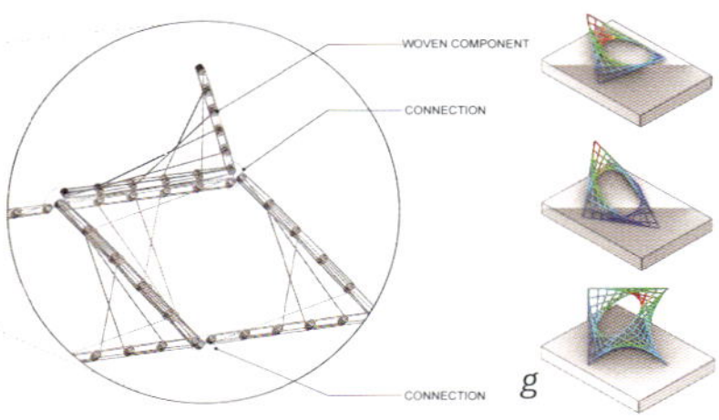


*g*

*Students: Shunta Moriuchi, Franklin Wu, Joey Luo, and Shane Su*

*The project proposes a method of semi-autonomous generative design for a garden canopy that incorporates a structural optimization method in combination with virtual building agents that are constrained to a filament-winding fabrication method.*

*BUILDING*: FALL SECTION II

*a) Photograph of robotically fabricated prototype, b-d) intial optimization of a space filling aggregate geometry, a-g) agent-based filament-winding operations, h,j) perspectives from below canopy, i) aerial view of canopy structure, and k) elevation of porous canopy structure.*

*Students: Chunze Li, Yidan Wang, Matt Ward, Amber Chen, and Sihan Li*

*A modular tetrahedral spaceframe is fabricated with a variable density and scale to operate as a climable, 3D garden pavilion. Density varies in relation to structural and planting considerations.*

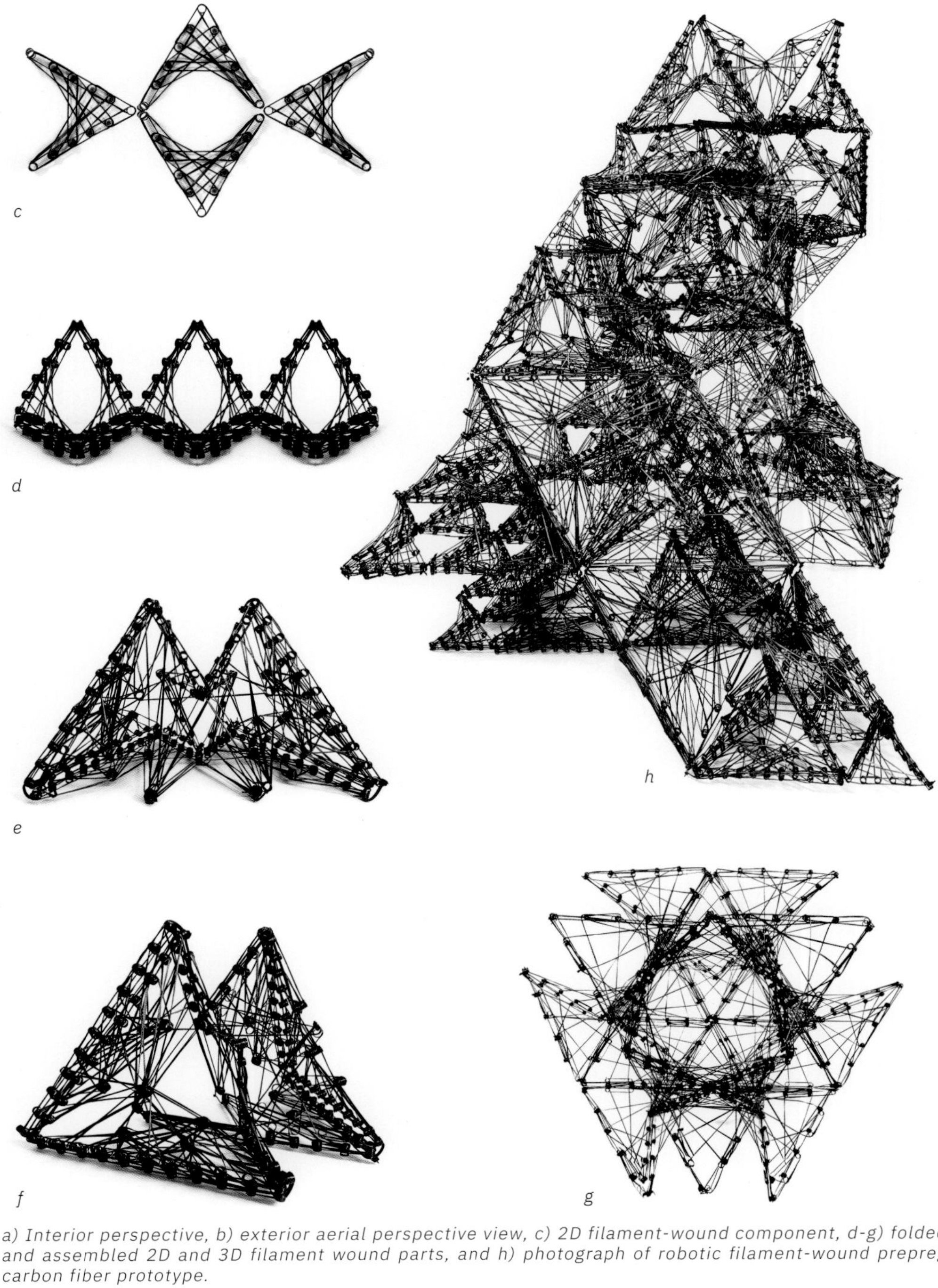


*a) Interior perspective, b) exterior aerial perspective view, c) 2D filament-wound component, d-g) folded and assembled 2D and 3D filament wound parts, and h) photograph of robotic filament-wound prepreg carbon fiber prototype.*

a

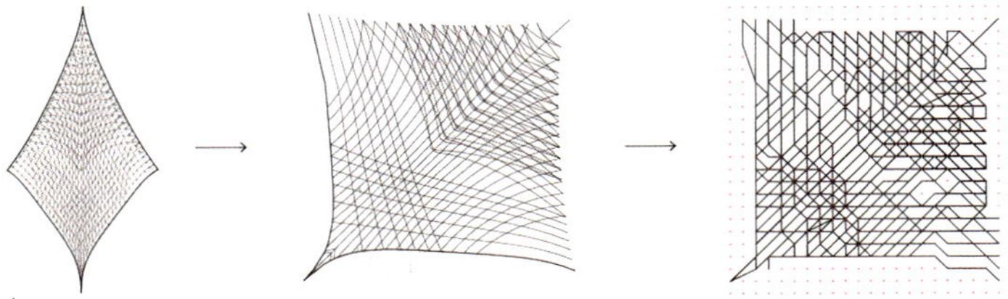

b

c

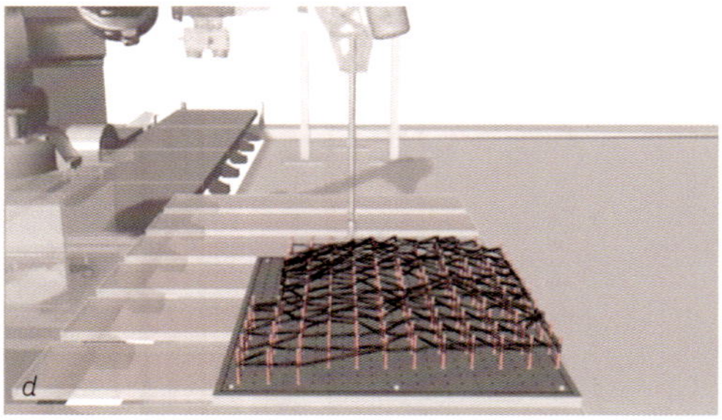

d

e

*Students: Cem Akgun, Aisha Alshehri, Adhityan Anbumozhi, and Adam Blood*

*A tensile surface concept is demonstrated for a filament-wound in prepreg carbon-fiber. A custom algorithmic method develops a weaving pattern that enables filament to be orientated to approximate both geometry and structural force within the surface.*

*BUILDING*: FALL SECTION II

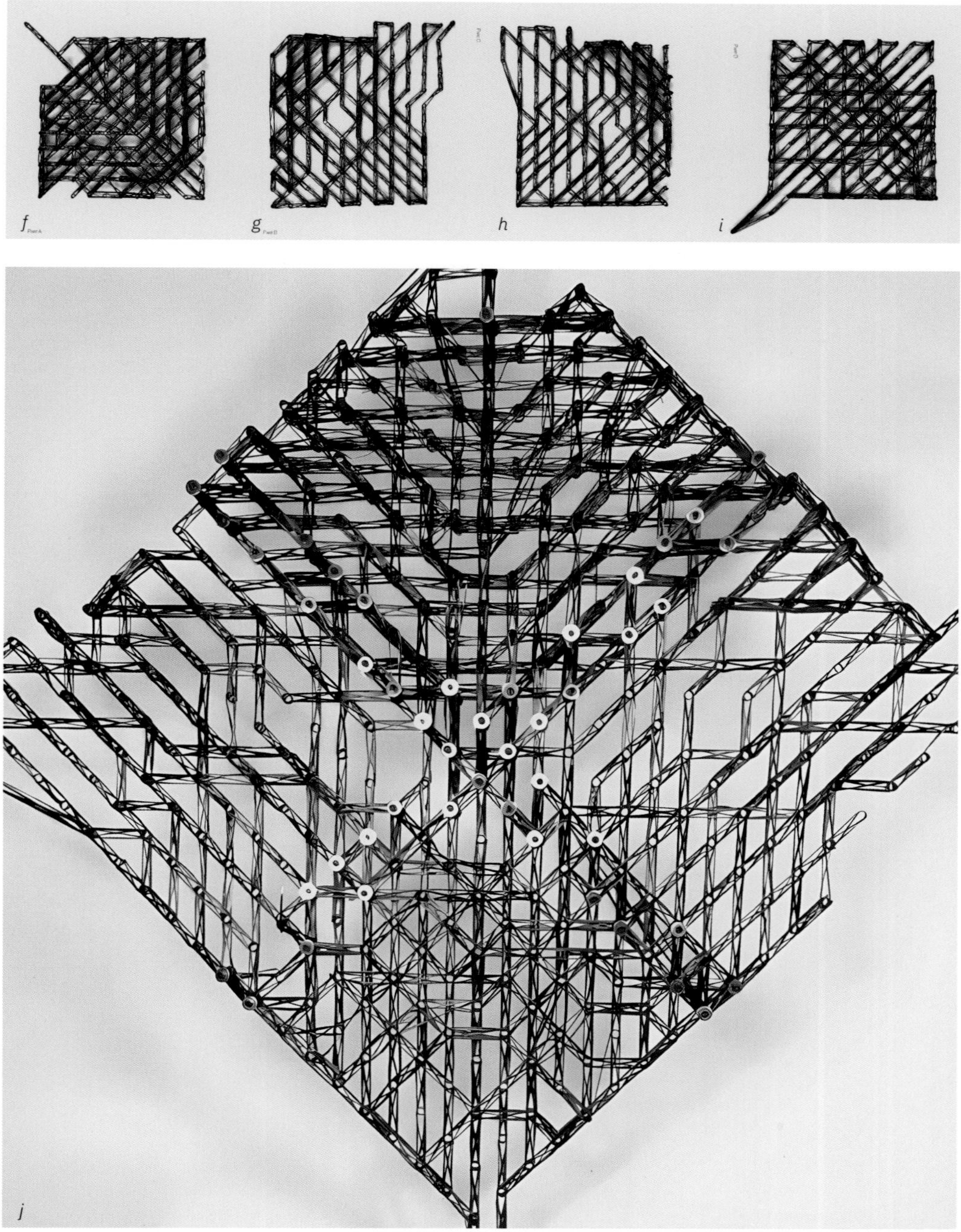

*a) Top roof plan of tensile installation, b) translation of tensile membrane to filament-winding pattern, c) rendered perspective, d-e) robotic filament-winding setup, f-i) four unique filament wound parts, and j) robotically fabricated prototypical assemblage of parts f-i.*

# REFLECTING
## on Spring Semester Curricula

*Faculty Deliberations*

# Integrating Design and Production:

## *Creative Approaches to Material Formation, Computation, Robot Control, and Tooling*

***— Robert Stuart-Smith***

**A**s the average beef consumer seldom has much knowledge of the abattoirs or butcher's methods, graduate architecture students are often unfamiliar with many of the industrial processes by which building components are manufactured (e.g., the difference between cold-formed and hot-rolled steel or injection-molded and die-cast extruded plastics). Perhaps this is partly due to a contingent of the profession that primarily conceives of buildings as an assemblage of existing industrial products, a view that was, to a degree, promoted by the second Director of the Bauhaus, Hannes Meyer (Meyer 1976). Architecture not constructed in situ does indeed consist of several industrially manufactured products ranging in scale from a brick to that of an insulating glass unit (IGU) or larger. Yet, this does not imply that architectural thinking should not engage with these constituent aspects of architecture in the same way that an architect might design precast concrete building elements. Moving beyond such undermining attitudes not only liberates design to strategically orchestrate more impactful production workflows but also creates a critical awareness of the ethical nature of industrial processes on which the industry is dependent. McDonough and Braungart's 2010 book *Cradle to Cradle: Remaking the Way We Make Things*, raised architects' awareness of the lifecycle and material make-up of buildings' constituent parts and advocated for cradle-to-cradle thinking (McDonough & Braungart 2010). Although after *Cradle to Cradle*, it is untenable for an architect to blindly specify a product known to be exceptionally harmful to the environment, there are several other ways architects can take a critical and creative approach to production. There persists a great need to reduce the environmental and economic cost of building without reducing the quality of our built environment or the degree of creativity embodied within its design. Only through more integral approaches to design and production can creativity be expanded in parallel to reductions in cost and environmental impact.

*Close-up image of robotically manipulated ceramics. An additively manufactured clay part is "smooshed" prior to bone-drying, and bisque and glaze firings (spring semester project: Robotics Prometheus).*

Architecture is rarely explored or strategized creatively as a form of material production. An architect's designs are typically developed and communicated through 3D Building Information Models (BIM) and drawings that are visualized through perspective drawing methods developed in the fifteenth century by Renaissance architects, including Brunelleschi, Alberti, and others. While a 3D BIM file provides an effective medium to communicate and reproduce construction-orientated intentions with associated material and product data, designs developed in BIM software primarily represent geometry. Such design workflows do little to expand the architect's thinking around materially formative manufacturing processes that could be instrumental in bridging design and production considerations holistically.

## Integrating Design and Production

New design freedoms have frequently emerged from the development of new production methods. In *The Projective Cast*, Robin Evans discusses how stereotomy was used both by stonemasons and architects such as Philibert De L'Orme in sixteenth-century France to design novel curve geometries that were realizable by stonecutting techniques (Evans 2000, 179-239). Pierre Luigi Nervi's magnificent isostatic reinforced concrete slabs' formal complexity was economically and practically possible due to his conception of a patented reusable mold system (Gargiani et al. 2016, 224-237). Jean Prouvé's career spanned from custom-developed furniture to mass-produced design-for-disassembly housing and other building systems. Prouvé's designs technically and aesthetically leveraged his knowledge and experience with metal fabrication processes (Lynch 2020). Such alignments between design and production can also extend beyond disciplinary design boundaries to further expand creative agency. Brunellesci's Grand Duomo cupola design was only realizable because of the designer's development of a new scaffolding system (Prager & Scaglia 2012). Although designers are not obliged to engage in such activities to develop progressive work, there are creative opportunities to be explored in the incorporation and experimentation of material production methods within design research. To explore these effectively, however, necessitates some knowledge of current industry practices.

Perceiving a gap in knowledge between industry and the architect, Walter Gropius, founder of the 1925 Bauhaus School, structured its pedagogy to unify the creative arts and bring them closer to industry. He saw this alignment as empowering, providing the designer with greater agency, and believed industry would value, seek out, and utilize graduates' expertise (Gropius 2003, 314). This alignment with industry was not an acquiescence but an intention to collaborate and help shape industrial design. Fast-forward

*To undertake research-led design into clay-based robotic fabrication, students embraced and became familiar with all aspects of ceramics production, including pugging clay bodies.*

100 years to today, and while mass-produced product designs do sometimes initiate custom production approaches to create economic value, the limited scope of a building's production seems often to undermine economic and strategic relationships between architecture and its industrial means of production. This is largely due to the fragmented nature of the real estate and construction markets.

The ideation and realization of a building typically involves a disparate organization of architects, engineers, general contractors, and sub-contractors. Together these groups form a project-based, ad-hoc construction partnership that is slow, inefficient, wasteful, and costly. In most cases, this loosely bound organization is also dismantled and built anew for every project, gaining little efficiency from repeat clients or projects that prevent it from operating at economies of scale, as product designs tend to do. This industry fragmentation also disincentivizes long-term thinking, with all parties' policies primarily operating on a project-by-project basis. Across several different contractual models, designs are frequently undermined. At one end of the spectrum, extensive geometric rationalization or value-engineering of a design might be required if a specific fabrication method is not considered at the outset or if a decision is made to change methods. At

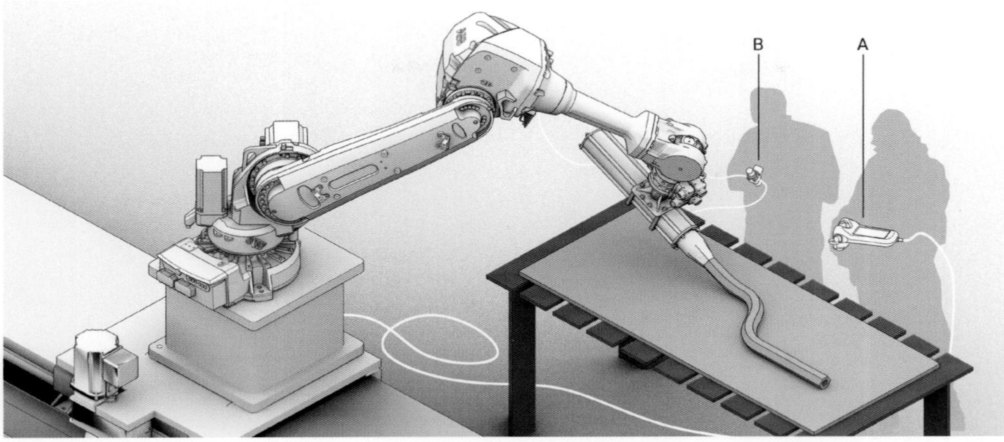


*Develped robot workcell with custom die attached to a pneumatically actuated plunger cylinder. Robot motion path programming enables variable extrusion paths to be achieved (spring semester project: Ceramic Forest).*

the opposite end, a design might be over-constrained by commencing with an industry manufacturing approach that is never questioned. This state of affairs is not sustainable and is currently being disrupted by a plethora of startup companies' augmented models of design, manufacturing, and construction. These intiatives are providing opportunities for greater alignment between the conceptualization and realization of architecture.

*Custom extrusion die additively manufactured in Onyx – a carbon fiber and nylon composite 3D print filament (spring project Filagree Façade).*

## Design for Manufacturing and Assembly

Attempts at integrating design and production considerations have been in development since early pursuits into mass production during the Second Industrial Revolution. In parallel to Frederic Winslow Taylor and Henry Ford's development of the assembly-line process, Ford's 1903 model-A car was designed to leverage assembly-line manufacturing processes (Goldense 2022). University of Massachusetts professors Poli, Graves, Murch, and Boothroyd began in the 1960s to develop a more generalizable set of principles that are now commonly referred to as "Design for Manufacturing and Assembly" or "DfMa" (Goldense 2022). DfMa principles ensure that individual parts are designed in consideration of their fabrication and assembly constraints and logistics, such as individual parts' maximum dimensions, orientation during fabrication, or production time-frames relative to an overall project schedule (Boothroyd 1994; Boothroyd & Alting 1992, 625-636). Some recent ventures are encouraging more architects to engage in DfMa principles by recasting architecture as a manufactured product. A non-exclusive sample of these ventures includes:

### *Volumetric Modular Prefabrication*

Volumetric modular construction is one approach to prefabricated architecture currently on the rise. Although the principles were championed more than a half-century ago in Archigram's Plugin City, Jean-Louis Chanéac's Cellules Polyvalent, Kisho Kurokawa's Nakagin Capsule Tower, and many other projects, today's methods are being demonstrated in larger-scale built work with far more agility and technical integration. For example, Forest City Ratner, Shop Architects, and Skanska's 2016, 461 Dean St project is comprised of prefabricated volumetric units that also operate as the building's structure. The project's 32 stories of units are stacked like Legos before being clad with prefabricated façade units (Touhey 2016). Despite substantial global interest and activity in volumetric modular construction, several modular construction companies recently dissolved (Obando 2021), suggesting that this approach does require some reevaluation and refinement. One could surmise that volumetric fabrication requires substantial factory floor space, equipment and employees that must be scaled to match increases in production, leaving a company financially extended if production output grows then periodically declines.

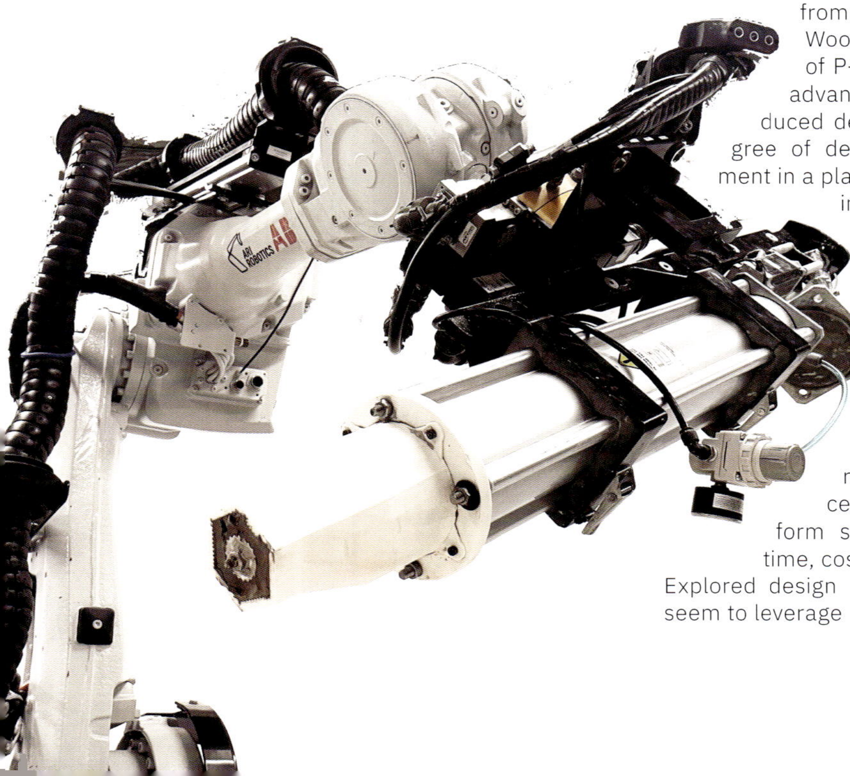


### *Robotic Prefabrication*

Several robotic timber fabrication companies have also emerged, such as Autovol and Blueprint Robotics, that produce prefabricated walls and elements, as well as volumetric modules, sometimes with integrated MEP services. Architects working with these fabricators are incentivized to collaborate early in the design process in order to align design constraints and opportunities to these companies' methods of manufacture. The Canadian company Intelligent City goes further, offering both design and manufacturing services for multi-story mass timber buildings. Intelligent City employs a parametric design model of a prefabricated building system and robotic manufacturing capabilities to enable the rapid and cost-competitive production of geometrically variable parts (Intelligent City n.d.).

### *Platform-Based DfMa (P-DfMa)*

Multi-disciplinary design and engineering company Bryden Wood, has developed a "platform-based" approach to building design and delivery for UK government projects that standardizes a building into a series of DfMa parts that can be used in diverse configurations to construct several different building typologies that integrate design and engineering input from several disciplines (Bryden Wood 2017). The standardization of P-DfMa parts offers economic advantages akin to mass-produced designs whilst enabling a degree of design specificity and investment in a platform together with variation in design outcomes. A pre-established pipeline for each part's manufacture simplifies fabrication and opens up possibilities for manufacturing to be distributed across several different companies in parallel if production demand necessitates such arrangements. Logistically successful, Bryden Wood's platform system offers reductions in time, cost and environmental impact. Explored design possibilities, however, don't seem to leverage robotic fabrication's capacity to support novel spatial, material, or aesthetic outcomes.

By designing in consideration of manufacturing constraints and workflows, companies such as Brydon Wood and Intelligent City have the potential to reduce waste and streamline fabrication and construction tasks. However, the standardization of architectural construction into too few a number or types of variably sized DfMa or P-DfMa parts could also potentially disincentivize experimentation and long-term evolution in design and production possibilities. There is also the potential that such standardized parts might partially homogenize the built environment, as similar to a building comprised of mass-produced parts, such systemic approaches pre-exist an individual project's specific considerations. However, every building site is unique in size, shape, character, use, and human occupancy to some degree, warranting more variable design and delivery methods. In contrast, emerging autonomous manufacturing technologies associated with the Fourth Industrial Revolution (IR4) are allowing designers to have greater control over production logistics, whilst enabling more uniquely specific design outcomes. IR4 supports a more diverse and sensitively designed built environment that is likely to generate a more abundant space for innovation moving forward, as can be seen already with additive manufacturing. Hybrid approaches to P-DfMa and additive manufacturing might offer additional benefits.

### *Additive Manufacturing*

Additive manufacturing is one IR4 technology that allows for substantial geometrical and aesthetic design freedom. The manufacturing method is relatively design agnostic – with a build cost primarily determined in relation to material volume and production time. Although this liberates aesthetic possibilities, most additively manufactured buildings realized to date by companies such as Icon (Dreith 2022) with on-site additive-manufactured concrete do not embody significant geometric inclination.[1] This is due to an absence of temporary support scaffolding and the concrete's material properties requiring curing time to gain sufficient strength to support several inclined layers above. In contrast to this, off-site additive manufacturing

*Above: 3D bundled façade screen prototype comprised of several geometrically unique die-cast extruded ceramic profiles. Left: Robot with custom die-cast extrusion end-effector tool (spring semester project: Ceramic Forest).*

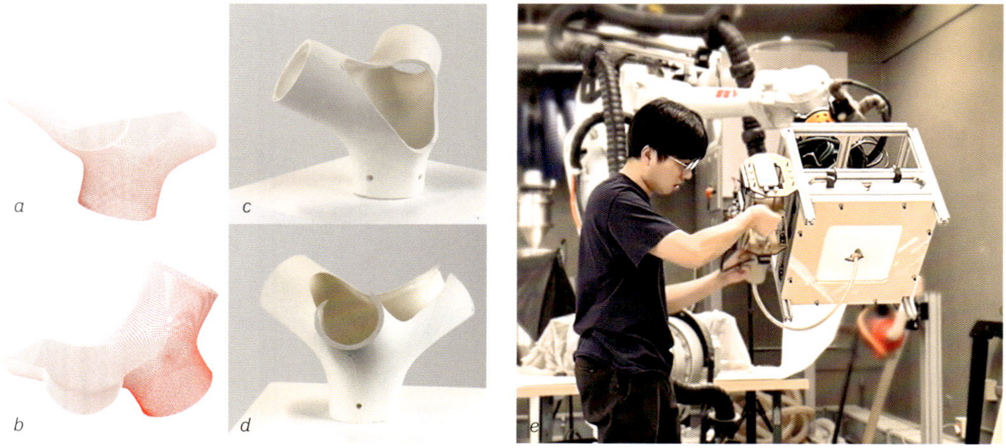

*Both simulation and hardware developments drive design possibilities in the spring semester Dynamic Slip-Casting project. a,b) Simulation timeframes during rotational slip casting, c,d) fabricated outcomes, and e) rotational slip-casting robot end-of-arm tool.*

companies are demonstrating more flexible geometric applications in several materials that rapidly harden, thus enabling more inclined geometries, supporting a greater diversity of design opportunities.

Companies such as MX3D and Nagami undertake both design and fabrication activities in-house, enabling more integrative thinking across these two logistically different activities. MX3D's wire-arc-additive manufactured (WAAM) metal pedestrian bridge in Amsterdam leveraged the possibilities of additive manufacturing to realize an unusually complex bridge design whose geometry's form and articulation align with structural forces within the bridge, enabling it to be quite materially efficient (MX3D 2020). Nagami's thermoplastic interior and exterior cladding projects and furniture are not structural yet explore other forms of material effects that are only possible through custom approaches to the additive manufacturing process (Jiminez-Garcia et al. 2023). Nagami develops its own design algorithms, robotic fabrication programs, and end-effector tools in-house and can leverage advances in these areas for aesthetic design benefit. While MX3D and Nagami's design outcomes explore exciting aesthetic possibilities, their aesthetic and production pursuits are entwined, enabling material and production-efficient design outcomes. Material waste and production efficiency affect their profit margins, incentivizing better practices.

Beyond additive manufacturing, there are many other manufacturing methods that can be augmented by robotic production methods, to enable greater variability and aesthetic inquiry. Such creative endeavors can be developed to demonstrate novel aesthetic outcomes whilst bridging longstanding gaps that have existed between architectural aesthetic intent, environmental ethics, economic viability, production efficiency, and more.

## Creative Manufacturing

The MSD-RAS curricula provide an integrated, holistic approach to learning that does not conceptually distinguish between creative and technical work. Instead, the program operates under the assumption that possibilities for innovation are at their best when these activities are considered together —particularly through critical and creative approaches to production. In the MSD-RAS, unique design outcomes are developed through an integrated approach to design and robotic manufacturing. In final spring semester projects, designs are developed first through research and development of a manufacturing approach. Design outcomes

a

b

*a,b) Post-processing of robotically fabricated slip-cast parts. Robotic slip-casting requires intimate knowledge in slip-casting and extensive fabrication tests to achieve a constant edge thickness (spring semester project: Dynamic Slip-Casting).*

seek to embody the manufacturing approach and to leverage it for creative and aesthetic expression. By introducing robotic control to produce variability within established industry manufacturing processes, many of these projects could theoretically be scaled to industrial volumes of production with varying degrees of time and cost efficiency.

In these projects, creativity operates in the programming of a robot's motion, the design of its end-of-arm tool (EOAT), sensing capabilities, and real-time feedback, or in the characteristics of a material, its methods of manufacture, or associated developments in a computational, parametric, or explicit digital modeling approaches to design. These all contribute to a research project's set of design intentions and constraints. Aesthetic expression is sought in these processes and honed in relation to production logistics to operate intrinsic to manufacturing activities. The alignment between aesthetics and methods of production erases any conflict between design conception and realization, streamlining the process while also enabling design agency to operate within production workflows.

The custom programming of robotic approaches to building manufacturing enables designs to be varied across parts or over several whole buildings, supporting greater specificity than mass production or mass customization manufacturing affords. More specifically,

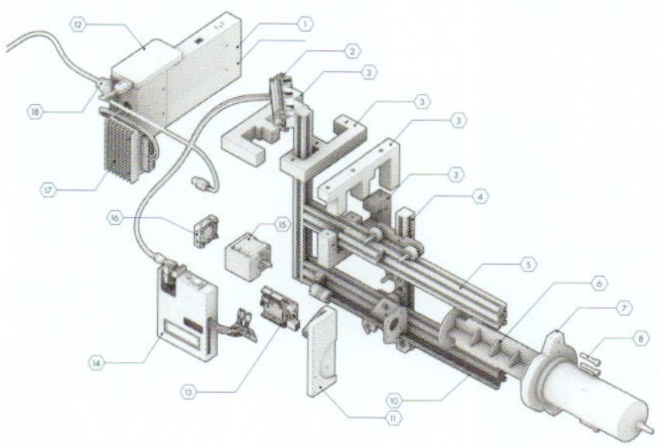
a

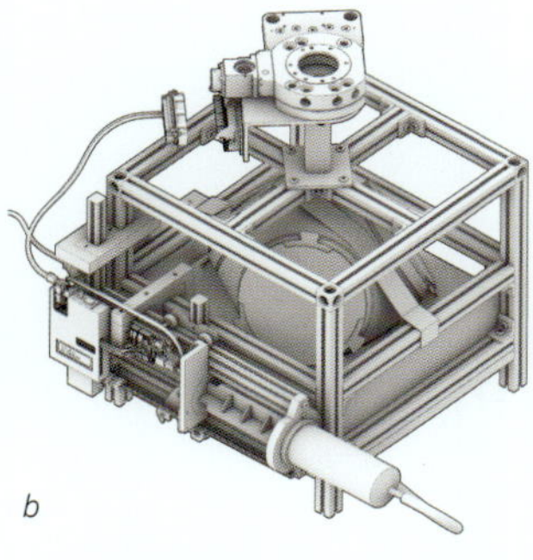
b

*a) Exploded slip micro-dosing mechansim and b) robot EOAT incorporating slip-cast mold and micro-dosing hardware (spring semester project: Dynamic Slip-Casting).*

computational design methods can also inform material processes; as architectural robotics pioneers Fabio Gramazio and Mattias Kohler argue, this enables a "digital materiality"

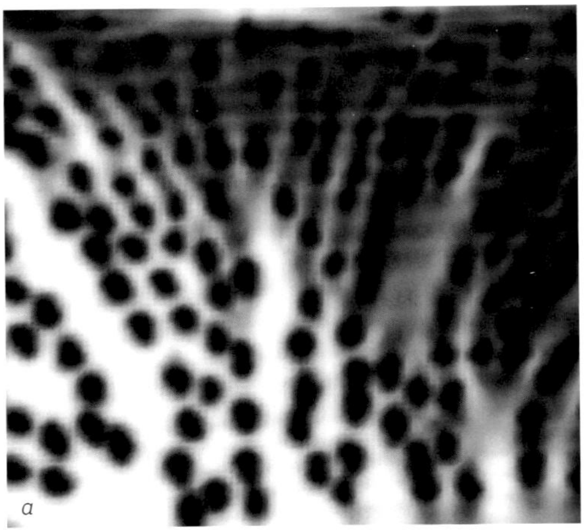

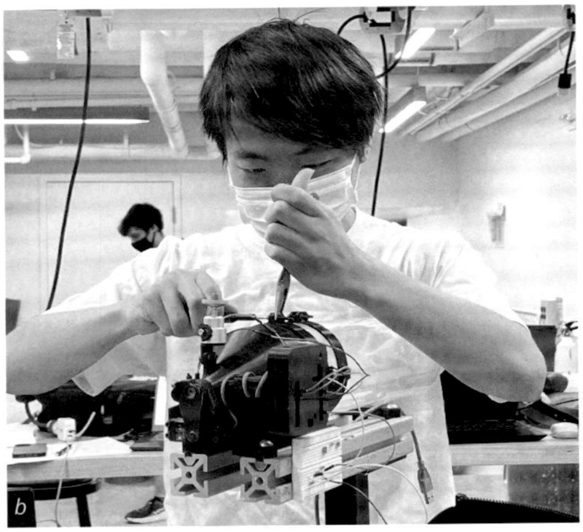

*a) Computational design method leveraging topological structural optimization (spring semester project: Sound Distortion) and b) student designed and fabricated robot end-of-arm tool (spring semester project: Hydroceramic Façade).*

offering new aesthetic and functional possibilities (Gramazio et al. 2014, 14). As discussed in the introduction, such robotic and computational approaches to design and production also potentially enable more sensitivity to diverse individuals, communities, environments, building typologies, species, and climates. Exploring integrated approaches to variable robotic fabrication and design is both an ethical and opportunistic stance, offering unprecedented potential in reducing environmental impact and cost, while increasing design quality and variability in geometric and material forms of aesthetic expression. This stance also places production and delivery sets of questions into the designers' field of perception, which can extend into questions of product lifecycle and end-of-life concerns. In this sense, the MSD-RAS program is, on the one hand, cognizant of industry developments, educating students for today's rapidly advancing industry, yet is also engaged in carefully considered, bespoke work that is driven more by speculative, curiosity-driven research and students' identification of gaps in research literature at the outset of their work.

In the MSD-RAS program, the role of the designer and their engagement with materials and means of production is challenged through varying degrees of design-production agency. These enable the designer/s to engage with the physical world in profoundly different ways. Robotic fabrication can extend production beyond automated routines with degrees of autonomy that enable sensitive material engagement to drive alternative forms of material expression. Beyond supporting bespoke or varied geometrical design outcomes, robotic fabrication enables the orchestration of custom manufacturing methods that allow the designer to influence materially formative processes. The designer becomes a contemporary artisan that operates almost vicariously through robot systems to offer an agile, industrially scalable approach to bespoke production.

Exploring such possibilities requires a change in mindset, where architects move from idealized, preconceived, and isolated activities equivalent to stage-performed ballet to the more adaptive and environmentally engaging thrill-seeking of parkour. Variable external conditions are recast as opportunities to develop hyper-specific creative outcomes.[2] This does not undermine the value of material or production-agnostic architectural ideas (nor classical ballet) but suggests there are possibilities to expand design's utility and creative inspiration by engaging with other material and productive forces.

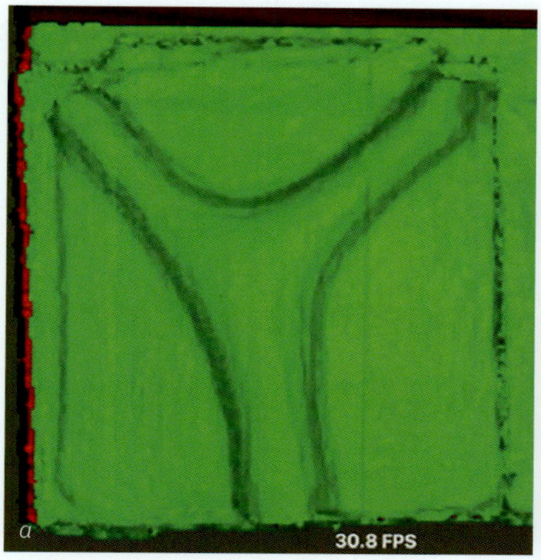


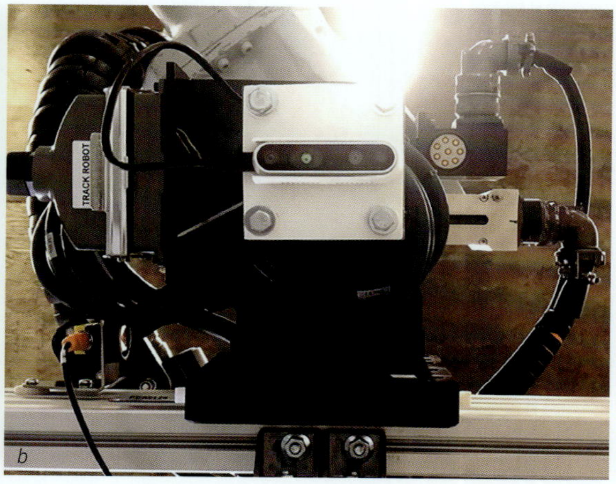


*a) 3D scan of incrementally formed part mid-way through fabrication and b) Intel Real Sense RGBD camera attached to robot EOAT for scanning of the part (spring semester project: Steamed Greens).*

## Embodied Computation

To operate in the physical world, one must perceptually and cognitively engage with it. Design for robotic fabrication is no different. While some fabrication tasks might be able to be executed sequentially and predictably, there are many forms of making that require iterative feedback between a maker and their material. In lieu of manual operation or offline (pre-determined) programming, industrial robots can also be programmed with degrees of autonomy. Semi-autonomous approaches such as behaviour-based robotics (Matarić et al. 2007) aim to render computation embodied within a robot platform and situated through sensor data to operate effectively in the world in an agile manner using online programming (Brooks 2002, 51–52). Computer vision and 3D scanning in-the-loop provide robots with autonomous perception and enable fabrication to respond to temporal material states. Although a sensor-packed end-effector tool is an appendage, it might also be seen as an extension of a robot platform's body and mind. From another point of view, some design activities undertaken in the MSD-RAS are experienced vicariously, through the sensing and orchestration of adaptive fabrication activities in other bodies.

Documented in this book, the first three-year research agenda of the program's final semester explored the development of robotic fabrication and design approaches for architectural façade screen ceramic assemblages. Clay's extensive architectural history and its vast number of industrial and artisanal forming methods provided fertile ground for this research. Given its plasticity, there were also exciting opportunities to leverage sensor feedback to dynamically engage in such formative processes robotically and creatively. Not only might a tool or robot motion path direct clay formation, but the development of underlying computational programming provided opportunities to address such operations at different scales of order concurrently, from a materially formative scale to that of a façade or building scale.

## AI and Computational Methods

In the MSD-RAS, software is written by students in several programming languages to enable variable and conditional acts to take place in tool or robot activity, simulation or design modeling, and other areas. Generative design approaches employing multi-agent systems or machine learning methods are explored alongside optimization tasks involving genetic algorithms or deep reinforcement learning. Computational processes are integral to design and fabrication methods and open up possibilities for defining these activities on unique terms rather than through reliance on generic established software solutions. While nascent large-language model (LLM) AIs such as ChatGPT (OpenAI et al. 2024) might soon be able to effectively program software on request, for now, design and production activities are curated by student programming initiatives. Programming is yet another frontier where technical and creative activities are inseparable, and knowledge in this area is likely to remain beneficial in LLM-enhanced industries for some time to come. In MSD-RAS final semester projects, software and programming languages operate similarly to a pencil, offering another means to communicate and realize architectural ideas. Inseparable from their aesthetic outcomes, these endeavors consist of several different forms of programming activities including:

*Simulation Modelling*
MSD-RAS final semester projects seek to render aesthetic outcomes as the sum of several contributing digital and material processes. Engaging with dynamic material processes often requires either live feedback on physical material states or accurate means to simulate such states prior to or during production. This is most necessary where a production process is not predictable or is visually inaccessible to the designer – such as the state of clay slip on the interior of a mold. In such cases, some form of fabrication simulation modeling (e.g., a particle-spring physics soft or rigid body simulation) becomes a means by which to both visualize and realize a design outcome.

*Generative Design & Machine Learning Processes*
Generative design algorithms are developed for other creative acts that operate prior to fabrication within a 3D modeling environment. To the unacquainted, computational methods' logical and procedural nature might suggest that they have no creative capacity. However, this depends on the way data and methods are associated with one another. While a linear set of associations found in a parametric model or simple algorithm might offer only variation within a predefined set of criteria, there are other algorithms that produce emergent outcomes that could not be ascertained or predicted from their rules alone. Emergence describes an observable, qualitative order or pattern that is not explicitly defined in the rules of a system. Computational approaches such as multi-agent systems, cellular automata, and other distributed, autonomous programming solutions offer one avenue for creative programming. Equally, deep learning approaches such as convolutional neural networks (CNNs),

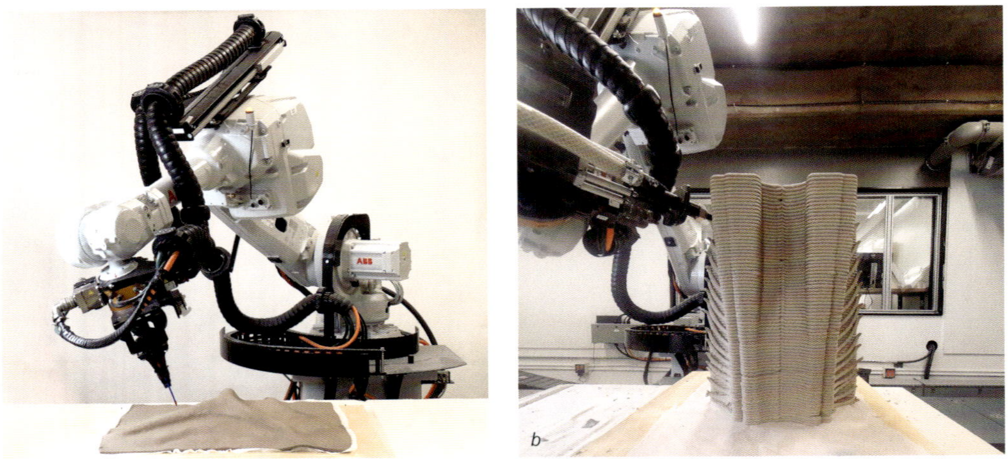

*a) Robotic relief carving (spring semester project: Steamed Greens) and b) 6-axis conformal additive manufacturing ontop of a previously additively manufactured clay part (spring semester project: Robotic Prometheus).*

a b c

*Studies of robotic smooshing of clay slabs prior to their bone-drying and bisque and glaze firing. Qualitative effects arise from both robotic operations and glazing finish (spring semester project: Robotic Prometheus).*

generative adversarial networks (GANs), and others develop unpredictable and complex design solutions from an extensive set of interconnected data associations. In the MSD-RAS, creative algorithms are leveraged in computational design methodologies where they can be developed to address specific architectural, robotic fabrication, and aesthetic design criteria, to produce creative outcomes that are not explicitly defined in their rules.

### *Deep Reinforcement Learning*

Deep reinforcement learning (RL) approaches are invaluable where a problem must be overcome that does not have a clear or small set of contributing factors, or where the solution might not be consistent under variable circumstances. An RL model might be used to train a design or robot fabrication task or a task that integrates both activities. Where a complex or variable task must be developed to work robustly, RL methods can produce solutions by iteratively refining action-planning in response to varied circumstances in a multitude of simulation or physical tests before being deployed in manufacturing tasks.

## Encouraging Collaboration and Innovation

An integrated approach to design and production extends the agency of the designer into other fields of research not traditionally located in an architectural scope of work, encouraging a collaborative space where diverse individuals can initiate and realize novel work that is the sum of all parties' knowledge and skills. Many of the world's most successful business leaders have likened innovation to the connecting of dots others haven't imagined could be connected. CPI Head of Marketing Communications, Matthew Herbert, suggests "integration and collaboration can be a systematic way to innovate" (Herbert 2019), while business catalyst and speaker Peter Fisk describes how entrepreneur Sir Richard Branson seeks to build Virgin's innovative and collaborative culture around Branson's life principle of "A+B+C+D (Always Be Connecting the Dots)" (Fisk 2016).

In the MSD-RAS program, students develop a wide range of dots, attaining knowledge and skills within a series of concurrently run courses by different faculty who have complementary knowledge and experience. As MSD-RAS's curricula and design projects are multidisciplinary in nature, the experience of developing these projects also encourages and empowers alumni to engage in more extensive multi-disciplinary collaborations post-study. For instance, while an MSD-RAS student can design and build an actuated robot tool with sensors, they are also well-equipped to collaborate with someone in the field of mechatronics who has more extensive knowledge in that area. Likewise, their knowledge and experience in applying computer vision to manufacturing and design applications makes them well-placed to collaborate with a computer scientist to extend vision-based manufacturing capabilities further. Equally, knowledge of ceramics or other materials enables a productive collaboration with an experienced and talented ceramicist. Attained knowledge empowers new collaborations and possibilities for new fields of enquiry, practice, and entrepreneurialism.

MSD-RAS's integrated approach to design and production aligns with Branson's advocation for always connecting dots. The program explores design beyond the sketchpad or 3D model and considers any technical or logistical ingredient from conception through to manufacture and assembly as part of the tableau or creative medium. Viewed from this perspective, MSD-RAS's final semester design-research projects seek to lay the groundwork for novel, innovative, and integrated approaches to design and production. Through such approaches, the MSD-RAS program opens up new avenues to collaborate and develop architectural designs within a greater cultural, environmental, and industrial context that becomes the substrate for novel, critical, and creative work.

## Notes

1. Concrete additive manufacturing using the continuous material extrusion chemical reaction bonding method (MEX-CRB).

2. External conditions here refers to material and production concerns that are traditionally outside of the remit of the architect. These can be engaged with when fabrication is incorporated into design research.

## References


Autovol. "Making Smart Construction Brilliant". Accessed on July 25, 2023. https://autovol.com/

Blueprint Robotics. "A Better Way to Build." Accessed on July 25, 2023. https://www.blueprint-robotics.com/

Boothroyd, G. "Product Design for Manufacture and Assembly." *Computer-Aided Design* 26, no.7 (1994): 505-520.https://doi.org/10.1016/0010-4485(94)90082-5

Boothroyd, G., and L. Alting. "Design got Assembly and Disassembly." CIRP Annals: Manufacturing Technology (italicized) 41, no. 2 (1992): 625-636. https://doi.org/10.1016/S0007-8506(07)63249-1

Brooks, R. *Flesh and Machines: How Robots Will Change Us*. New York: Pantheon Books, 2002.

Bryden Wood. *Delivery Platforms for Government Assets: Creating a Marketplace for Manufactured Spaces*. London: Bryden Wood Technology, 2017.

Dreith, B. "Construction Commences on BIG and ICON's Community of 3D-Printed Homes in Texas." Dezeen. November 11, 2022. https://www.dezeen.com/2022/11/11/icon-big-wolf-ranch-3d-printed-homes-austin/

Evans, R. *The Projective Cast: Architecture and Its Three Geometries*. Cambridge: MIT Press, 2000.

Fisk, P. "What is Innovation? Connecting the Dots: The Ones Other People Miss." Peter Fisk. February 26, 2016. https://www.peterfisk.com/2016/02/what-is-innovation-connecting-the-dots-the-ones-other-people-miss/

Gargiani, R., A. Bologna and J. Haydock. *The Rhetoric of Pier Luigi Nervi: Concrete and Ferrocement Forms*. Boca Raton: CRC Press LLC, 2016.

Goldense, B. L. "A History of Design for Manufacturing and Assembly." Machine Design. January 7, 2022. https://www.machinedesign.com/automation-iiot/article/21213546/a-history-of-design-for-manufacturing-and-assembly

Gramazio, F., M. Kohler and J. Willmann. "Authoring Robotic Processes." *Architectural Design* 84, no. 3 (2014): 14-21.

Gropius, W. "The Theory and Organization of the Bauhaus." In *Art in Theory 1900-2000: An Anthology of Changing Ideas*, edited by C. Harrison and P.J. Woods. New York: Wiley, 2003: 309-314.

Herbert, M. "Connecting the Dots for Innovation." Centre for Process Innovation Limited. November 9, 2019. https://www.uk-cpi.com/blog/connecting-the-dots-for-innovation

Intelligent City. (n.d.). "Adaptive Mass Timber Automation Becomes a Reality at Intelligent City." Accessed October 29, 2023. https://intelligent-city.com/announcements/adaptive-automation-becomes-a-reality-at-intelligent-city/

Jiminez-Garcia, M., I. Viguera Ochoa and M.Á. Jiménez García. "Projects" Nagami. Accessed on May 4th 2024. https://nagami.design/en/projects/

Lynch, P. "Spotlight: Jean Prouvé." ArchDaily. April 8, 2020. https://www.archdaily.com/785158/spotlight-jean-prouve

Matarić, M. J., N. Koenig and R.C. Arkin. *The Robotics Primer*. Cambridge: Cambridge University Press, 2007.

McDonough, W., and M. Braungart. *Cradle to Cradle: Remaking the Way We Make Things*. New York: Farrar, Straus and Giroux, 2010.

Meyer, H. "Programs and Manifestos on 20th Century Architecture." *The Journal of Aesthetics and Art Criticism* 34, no. 4 (1976): 117-118. https://doi.org/10.2307/430599

MX3D. "MX3D Bridge." MX3D. 2020. Accessed on May 3rd 2024. https://mx3d.com/industries/mx3d-bridge/.

Obando, S. "What Does Katerra's Demise Mean For the Contech and Modular Industries?" Construction Dive. October 13, 2021. https://www.constructiondive.com/news/what-does-katerras-demise-mean-for-the-contech-and-modular-industries/608037/

OpenAI, Achiam, J., Adler, S., Agarwal, S., Ahmad, L., Akkaya, I., Aleman, F. L., Almeida, D., Altenschmidt, J., Altman, S., Anadkat, S., Avila, R., Babuschkin, I., Balaji, S., Balcom, V., Baltescu, P., Bao, H., Bavarian, M., Belgum, J., ... Zoph, B. (2024). GPT-4 Technical Report. https://arxiv.org/abs/2303.08774

Prager, F. D., and G. Scaglia. *Brunelleschi: Studies of His Technology and Inventions*. Mineola: Dover Publications, 2012.

Touhey, M. "SHoP Tops Out World's Tallest Modular Building in Brooklyn." DesignBoom. 2016. https://www.designboom.com/architecture/shop-architects-461-dean-street-pacific-park-brooklyn-new-york-05-15-2016/

*Robotic smooshing of an additively manufactured clay part. Qualitative effects arise from both robotic operations on the material and from its glazed finish (spring semester project: Robotic Prometheus).*

# COLLABORATING WITH AUTONOMOUS SYSTEMS

*— Jeffrey S. Anderson*

Three students stand hunched over a robotic work area, focusing intently on an ABB IRB-120 robotic arm slowly weaving its way through a metal jig. They are eager and attentive, but also weary. They have clearly done this before and they know that the greater their focus now, the better their results will be. As the end of their fall semester nears, they know that they need to get this digital fabrication project right. One student slowly feeds carbon fiber into the robot's end effector, carefully holding the appropriate tension to keep the fiber taught without bending the jig. Another student physically holds the jig while the robot slowly weaves through it, focusing intently on the end effector, ready to shout "stop" in the event of a collision, tangle, or loss of tension. The third student holds the robot's teach-pendant, her fingers gingerly hovering over the pause button. A bead of sweat forms on one student's brow. The robot bumps into the jig. "Stop!" The robot comes to a halt. The students jump into action, hurriedly making minor corrections to the jig, tightening the carbon fiber that has already been deposited, and adjusting the fiber spool. "Okay!" The robot continues its toil and the students continue their own. I have observed this form of human-robot collaboration many times in architecture schools, and believe it offers us an opportunity to explore the roles of humans and robots in an imminent era of pervasive human-robot interactions. We can use this example to be critical of the nature of our collaboration with robots and other autonomous systems but also to investigate what it means

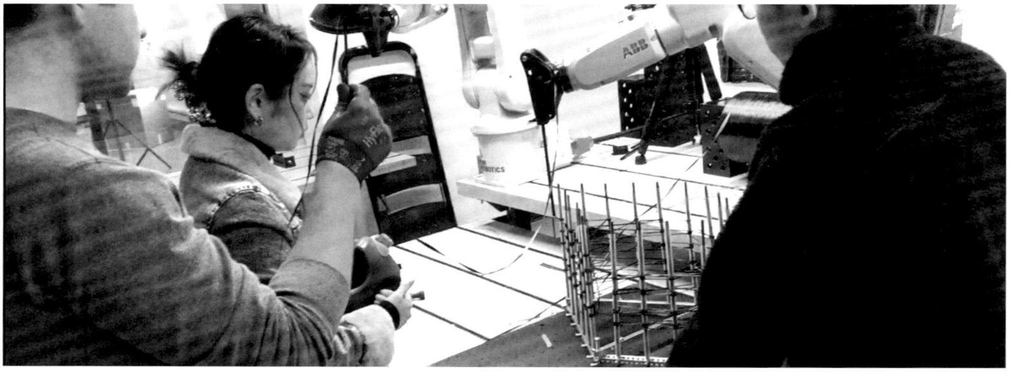

*A group of students hunched over a robot, each undertaking laborious work, keeping pace with the robot's operations.*

to be a human at a moment when computers are increasingly able to behave autonomously.

Collaborative robots or "co-bots" are a new generation of robotic arms that are designed to work safely and closely with humans in a shared workspace. In a traditional robotic workflow, humans are isolated from robotic work areas for safety reasons, typically separated with cages or other barriers because the robots are for the most part unaware of their surroundings. However, because co-bots are equipped with collision sensors and other safety features that cause them to shut down upon detecting unplanned contact, they are capable of closely collaborating with humans without the risk of injury.

This type of robot, though typically not as strong, fast, or precise as their "blind" counterparts, is increasingly being used in light manufacturing applications. With software that allows them to be physically trained by humans with no knowledge of robotic programming, these robots are incredibly nimble and can be quickly retrained to do a number of different tasks in succession. In scenarios where a co-bot must be programmed to operate at a number of different workstations, they need a collaborative relationship with a human operator to physically move them and reprogram them. This has begun to alleviate the burden of repetitive tasks from humans in light manufacturing, allowing increased production for small businesses.

An important component of the 4th Industrial Revolution is the ability for industrial production systems to operate with increased autonomy. Co-bots, through their ability to sense and interact with the world, will be a contributor to this new regime in small scale and bespoke production applications. As robots take on increasingly important roles in industrial production, often working closely with humans, we should be critical of the nature of their collaboration. If we return to the anecdote of the hunched over students, we can see that the students have set up a collaboration which leverages several strengths of the robotic arm but does not consider the role of the human in a meaningful way. While the robot utilizes its precision, strength, and patience, the humans are tasked with overcoming the technical limitations of the

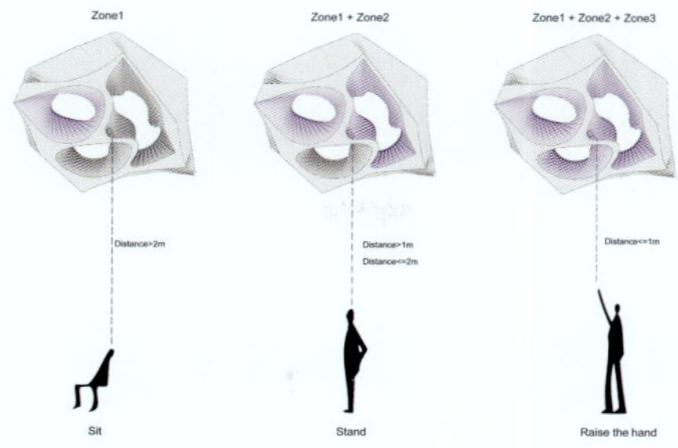


*a*

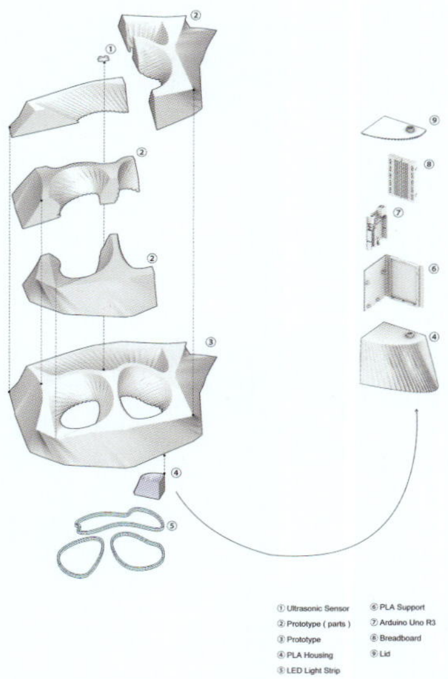


*b*

*a) Responsive lighting system responds to human activity determined by spatial proximity, b) exploded view of prototype, and c) a live demonstration of responsive lighting in the final prototype.*

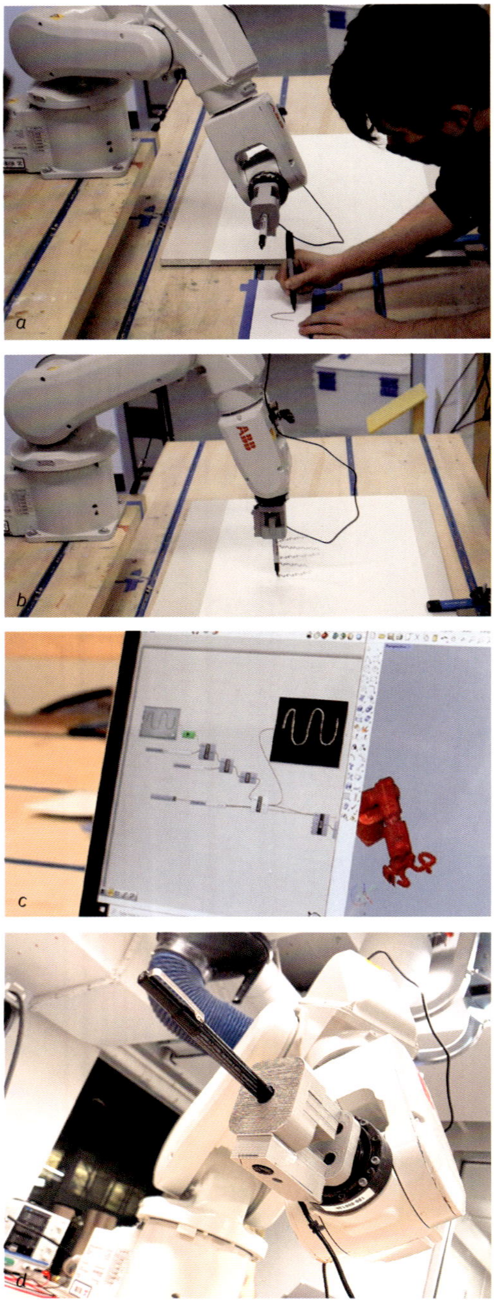

*a) Robot observation of human sketch, b)robot sketching in response, c) robot program development in Rhino3D Grasshopper, and d) robot EOAT with integrated webcam and pen.*

robotic arm, the jig design, and the end effector design. Humans engage in rote, repetitive tasks and roles that require physical endurance, such as holding the jig. Deciding when it is time to stop the robot and adjust the jig could be thought of as a higher level design role, however, jig holder, string feeder, and knot tier are not activities that set up humans in a meaningful collaboration role.

Obviously, this scenario is a stop-gap measure for students to complete a design project in a limited time frame, but it is a telling example of humans solving problems of their own creation. While the object the students were fabricating was designed around the limitations and abilities of the robot, new roles were created for humans in a technical ecosystem not optimized for either robot fabrication or human collaboration. If the jig was better designed and more securely attached to the workstation, or better calibrated to the robots coordinate system, and if the end effector itself was able to keep the appropriate tension on the carbon fiber, the fabrication process could be more autonomous. If these improvements had been in place, the students could have considered a more meaningful and creative collaboration, one in which they could have agency in the design process while it unfolds.

In a cultural moment defined by increasing automation, generative artificial intelligence, and big data, it is important to consider the role of the human in the production process and to reflect on how we should be spending our time in a world saturated with autonomous systems. An ideal scenario might be where humans collaborate creatively and in meaningful ways with robots while being mindful of, and utilizing the best capabilities of the machine. Alternatively, in a less desirable scenario, humans are obliged to have a collaborative role in which they have to overcome the technical limitations of the robot by completing mundane tasks in the service of increased industrial production.

In the Masters of Science in Design: Robotics and Autonomous Systems (MSD-RAS) program we utilize a wide range of robot platforms mostly consisting of ABB and UR robotic arms. While the ABB models are not technically co-bots, we equip them with sensing technologies that allow them to function like co-bots to some degree. In the spring semester 804 Advanced RAS Programming course, students explore human-in-the-loop fabrication scenarios, prototype speculative autonomous workflows, and demonstrate novel interactive systems with robotic arms. Over the last three years, several assignments build skills in robot programming and help students to develop critical thinking around human-robot collaborations. Students have engaged with a number of Extended Reality (XR) technologies to prototype these interactions including Virtual Reality (VR) equipment (HTC Vive and Meta Quest platforms), depth scanning cameras (Intel Realsense), microcontroller platforms (Arduino and Raspberry Pi), Augmented Reality (AR) technologies (AR Core and ARKit), and a variety of other sensors. In the fall 805 Introduction to Cyberphysical Systems course, students are introduced to programming concepts and are challenged to assemble electronic circuits using sensors, actuators, and microcontrollers, and integrate them into architectural objects. Building on this, the spring 804 Advanced RAS Programming course teaches a variety of skills that students can apply to their final research thesis project, that is realized within four courses that operate in different yet complementary subject areas.

The topics and assignments developed for this class are based on research, workshops, and exhibitions conducted over the last several years. In order to achieve real-time control in a variety of CAD/CAM environments, students use Robot Ex Machina, an open-source computational framework of software tools (García del Castillo y López 2019). Our AR/VR development environment is built in Unity3D and is aimed at providing a flexible and easily calibrated digital and physical environment for human machine collaboration (Johns et al. 2019). Students explore creativity and improvisation in robotic fabrication, setting up systems which allow them to make informed decisions on interactive robotic fabrication tasks (Johns et al. 2014) and they consider the role of the human in complex human-in-the-loop collaborative frameworks which overcome robotic shortcomings (Lafreniere et al. 2016). Exercises in this class encourage students to explore challenges related to the scanning and stacking of objects previously unknown to the robotic arm (Furrer et al. 2017) and recognition and grasping of irregular objects (Correll et al. 2016). The overarching principle guiding all of these assignments is a reframing of the role of robotics in human machine collaboration away from the linear execution of human design intent and towards a system that encourages true collaboration (Johns et al. 2018).

The Advanced RAS Programming course's assignments build up a series of increasingly sophisticated human-robot collaborative capabilities over the spring semester. In an

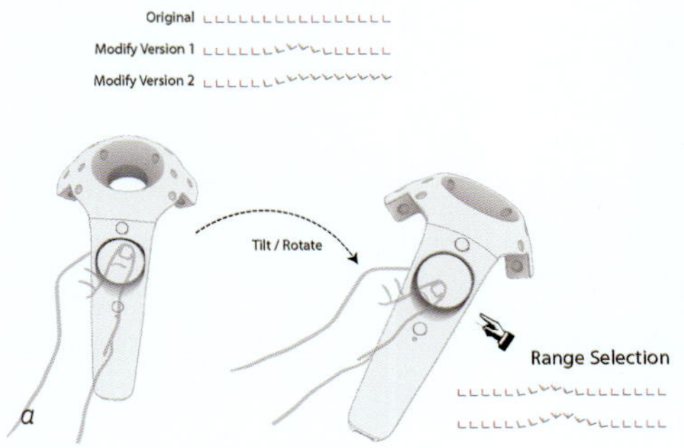

*a) Custom HTC Vive controller functions modify position and orrientation of an ABB robot's tool path, b) HTC Vive Controller in live control of additive manufacturing images shown on the next page ("Robotic Prometheus" project, spring semester).*

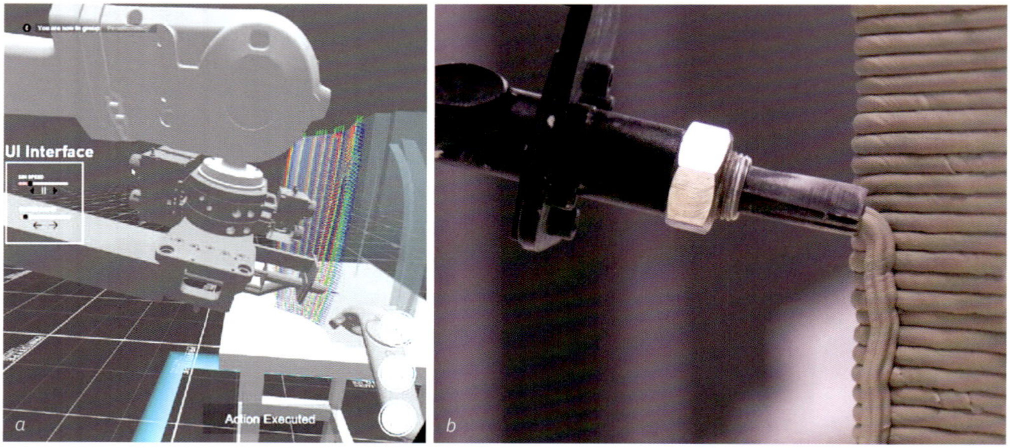


*a) Unity digital twin virtual model of robot work cell with tool path live updated from HTC Vive controller, b) ABB articulated robot arm additively manufacturing in clay with the live updated tool path. Conformal printing is challenging on irregular and vertical surfaces. Human updates helped fine-tune printing orientation in relation to material viscosity and the variable profiling of the existing clay part. Robotics Prometheus project in spring Semester.*

initial assignment, students complete an Interactive Drawing, using computer vision and interpretation to develop interactive drawing scenarios with a robot equipped with a webcam and a pen tool. The robot work area has a vision zone and a drawing zone. Students draw a sketch not previously seen by the robot, the robot then interprets this sketch and redraws it in its drawing zone. The robot can be programmed to manipulate or add details to the sketch such as changing its scale, adding additional shapes along its trajectory, or repeating patterns. In this assignment, students speculate on a collaboration in which a robot assists or augments a human's creative capacity, completing or adding detail to a simple sketch with robotic precision.

To introduce students to increased forms of robotic autonomy in fabrication scenarios, in a Sense and Respond assignment, students use a 1D depth sensor attached to a robot end effector to make changes to its tool path as it is running. This allows the robot to respond to unpredictable stimuli such as the degree of depth change while depositing clay on a flexible fabric. In this assignment, students speculate on a fully autonomous fabrication routine with unpredictable material outcomes, thus giving the robot the ability to respond to stimuli and make corrections to its toolpath without human intervention.

Building upon the previous interactions, a Depth Scan Deposition assignment tasks students with using an Intel Realsense depth camera attached to a robot end effector to scan a design-target geometry and a fabrication geometry while the robot is running. The robot is programmed to analyze differences between the two geometries and deposit material on the fabrication geometry, making them increasingly similar. In this case, students think of the robot as a director of material flows with which a design goal can be achieved without extensive pre-programming of its toolpath by using sensing technologies. In this assignment, students speculate on a playful and creative human-robot collaboration scenario in which a human designer can guide precise robotic fabrication behaviors using a safe and malleable analogous physical material.

In an Adaptive Pick and Place assignment, students program a robot to 3D scan and interpret a series of unknown objects and determine the order in which to place them in an ongoing assembly routine. In this assignment, students speculate on a collaborative fabrication scenario in which a human can provide a number of target objects to a robot without needing to sort or arrange them carefully and the robot can respond to these objects and make decisions about how to assemble them. This relieves the human collaborator of needing to do repetitive and precise operations.

As an introduction to XR technologies, students also complete an assignment based on Virtual Reality Tele-Operation. Students use Unity 3D, an HTC Vive VR headset, and a robotic arm to view, simulate, and manipulate a robotic toolpath while it is operating. By streaming the robot's joint angles and toolpath to a user in VR, a human can view the robot's progress from angles that would be unsafe in real life, preview robot motion with high precision, and engage with the robot in new ways by changing their scale, position, or environment. In this assignment, students speculate on a digital twin scenario in which a human technician can step into the robotic work cell virtually while the robot is operating, view real time fabrication conditions with point clouds streamed from the work area, make minor corrections to robotic toolpaths, and have the robot execute those corrections without stopping the fabrication process.

Altogether, in these assignments students gain experience in a number of human-robot collaboration scenarios and allows them to speculate on what automated systems should and should not do. Automated systems should augment human capacities and experiences, not replace them. We can develop human-robot collaborations that prioritize human creative expression rather than optimization, we can use computational systems to improve human decision-making abilities, and we can use technologies that improve the physical or mental capacities of humans to improve their lives. Automated systems should inspire and cultivate creative making and playful invention in humans, not replace it. We should not use tools like Chat GPT to write a love letter for us, we should use them to inspire, revise, and give suggestions to us so that we can undertake personal and meaningful tasks with fewer burdens.

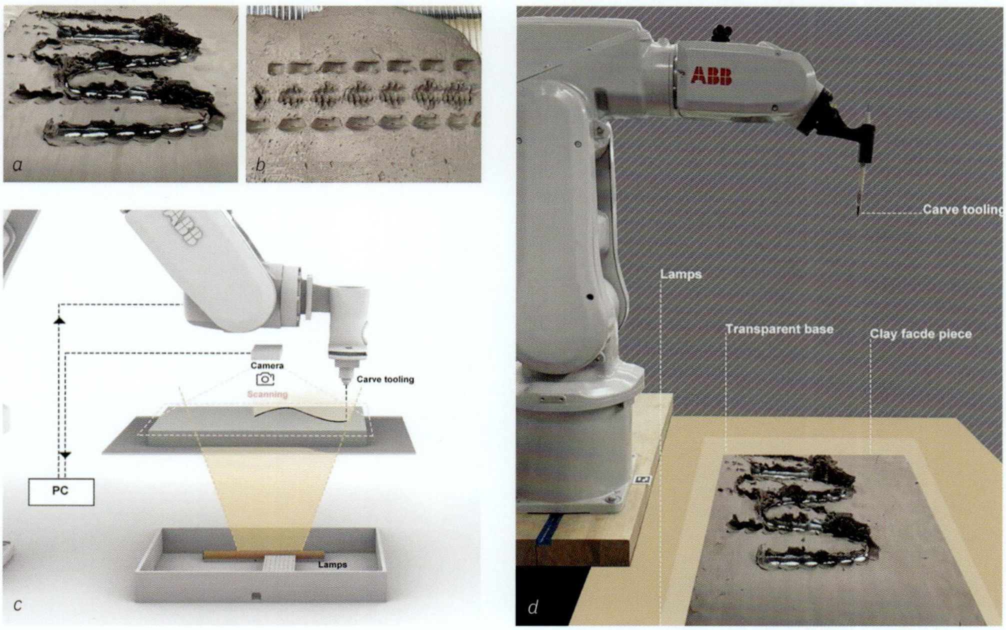


*a,b) Two carved outcomes using, c) a light-adaptive manufacturing work-cell that utilizes a webcam and computer vision to identify bottom-lit regions in the clay from perforations, and d) a robot workcell with carving tool. For image credits, refer to Light-ing project in spring Semester.*

It is the act of creative making itself that makes us human, not the quality or quantity of the content produced. Automated systems should help us produce and consume in a smarter way, not just produce more. They should give individuals or small businesses the capacity to grow and compete with powerful industrial agents, increasing their ability to create novel and unique outputs.

MSD-RAS graduates venture into the professional world equipped with a critical eye around how to engage with autonomous systems. After working with varying degrees of collaboration and autonomy in robot systems throughout their studies, they should have strong opinions about what kinds of automated systems they want to work with and how these can make human lives easier and more fulfilling. Those hunched over students, having toiled with their jig and robotic arm, should be able to conceive of a digital fabrication process in which they either program themselves out of the loop or program themselves into a meaningful creative collaboration. Given the imminent pervasiveness of autonomous systems, such as robotics and generative artificial intelligence, we must consider how these technologies can provide new opportunities for creativity and human engagement rather than serve as engines for ever greater content creation and content consumption.

## References


Correll, Nikolaus, Kostas E. Bekris, Dmitry Berenson, Oliver Brock, Albert Causo, Kris Hauser, Kei Okada, Alberto Rodriguez, Joseph M. Romano, and Peter R. Wurman. "Analysis and Observations from the First Amazon Picking Challenge." ArXiv. January 2016. http://arxiv.org/abs/1601.05484.

Furrer, F., M. Wermelinger, H. Yoshida, F. Gramazio, M. Kohler, R. Siegwart, and M. Hutter. 2017. "Autonomous Robotic Stone Stacking with Online Next Best Object Target Pose Planning." *IEEE International Conference on Robotics and Automation (ICRA)*, Singapore, January 2017, 2350-2356. https://doi.org/10.1109/ICRA.2017.7989272

García del Castillo y López, J.L. "Machina.NET: A Library for Programming and Real-Time Control of Industrial Robots." Journal of Open Research Software 7, no. 1 (2019): 27.

Johns R.L., , J. Anderson, and A. Kilian. "Robo-Stim: Modes of Human Robot Collaboration for Design Exploration." In *Impact: Design with All Senses, Proceedings of the Design Modelling Symposium*, Berlin, September 2020, 671-684.

Johns R.L., and J. Anderson. "Interfaces for Adaptive Assembly." In *Acadia 2018 Recalibration: On Imprecision and Infidelity, Proceedings of the 38th Annual Conference of the Association for Computer Aided Design in Architecture*, Mexico City, October 2018, 126-135.

Johns, R., A. Kilian, and N. Foley. "Design Approaches Through Augmented Materiality and Embodied Computation." In *Robotic Fabrication in Architecture, Art and Design*, edited by W. McGee and M. Ponce de Leon, 319-332. New York: Springer, 2014.

Lafreniere, B., T. Grossman, F. Anderson. J. Matejka, et al. "Crowdsourced Fabrication." In *UIST 2016 Conference Proceedings: ACM Symposium on User Interface Software & Technology*, Tokyo, October 2016, 15-28.

# PRECISELY IMPRECISE:

## *Ceramic Material Systems as a Framework for Design Robotics*

*— Nathan King*

This essay accompanies a compendium of the first three years of the Weitzman School of Design's Master of Science in Design: Robotics and Autonomous Systems program. During this time, the spring semester of the program has been characterized by work that explored the convergence of clay-based ceramics and industrial robotic fabrication possibilities. Reflecting on this work and its place within the field of architectural robotics research brings to mind an early venture into the manipulation of clay with an industrial robot some 15 years ago. At the time, we articulated a counter position to architectural robotics research, which focused on either engineering-oriented automation or one-off bespoke creative works. "Design Robotics" was coined to describe an alternative design research methodology that combined analytical research with creative discovery derived from iterative material-based experimentation. In contrast to the previous two approaches, Design Robotics intended to build a bridge between design innovation and industrial means of production (Bechthold & King 2013), where clay-based ceramics served as the vehicle by which we attempted to do weird, unconventional things with robots (Bechthold et al. 2015). Fast-forward 15 years to the present and I am still not an expert roboticist, ceramicist, or potter, relative to those who have dedicated their careers to one of these areas—nor are our MSD-RAS students. For the work we all do, the term Design Robotics seems even more relevant today in describing the illusion of subject matter expertise or perhaps the reality of having expertise that lies in between—an expert in understanding more to know less. That expertise in the between might

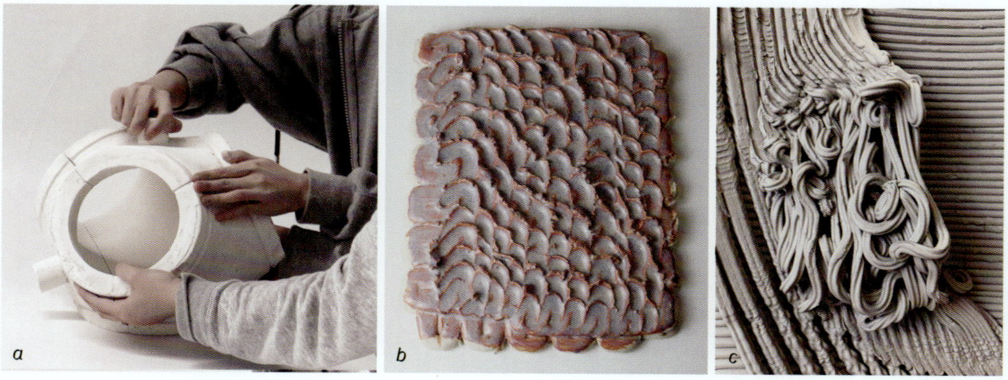


*a) Slip-casting requires considerable mold-making and ceramics knowledge, b) robotic clay smooshing, and c) additively manufactured clay is deposited vertically against an already manufactured volumetric part in such a manner as to produce gravitational-material effects. See spring projects a) Dynamic Cast, and b,c) Robotic Prometheus.*

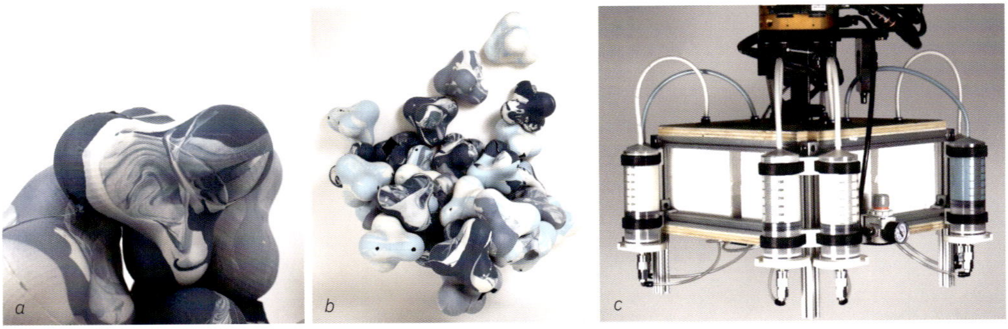

*a,b) Rotational robotic slip-casting produces unique marbling effects in each part that is intrinsic to a custom robot motion path sequence, considerable control of slip viscosity, and development of a custom tool c) which injects a specific ratio of two differently colored slips during fabrication. See spring project, These Things Swirl.*

be characterized as having an enhanced critical thinking ability which arguably is the value of design education at large.

There is an inherent tension that lies within Design Robotics between digital and material practices. The perceived precision of computation and digitally controlled tools leans toward the sanitary and intellectual. Conversely, the perceived sloppy and visceral nature of clay seems so far removed from that computational context and is, in some ways, alien to our digitally native students. However, the same can be true within a digital model where coincidence (or lack) of lines and curves, deviations amongst control points, surfaces, and plane orientations, or conflict in units and tolerances reveals a sanitized chaos that simply becomes sloppiness when that digital information becomes physically actionable. In a craft-based community ceramics studio there are often hundreds of people, dozens of clay bodies with many variations, and thousands of glaze compositions—all interacting in the same space without compromising one another. We see operational precision in an environment where "bowl 1" and "bowl 2" in a series of 12 bowls are similar but have variation—precise variation within a carefully controlled framework of temperature, moisture content, clay body, and skill developed over decades. In both digital and clay-based ecosystems there exists a world of tolerances. Our increasing ability to navigate, make decisions, and execute within those tolerances represents skill and craft. Navigating tolerances between diverse perceptions and realities in a way that arrives at an understood physical output is "practice" for each student to learn their own way of practice is also a goal of design education.

In the coincidence and contradictions of digital and material practices, we currently are experiencing a time-bound convergence in disciplinary thinking. Convergence is a beautiful idea that exists in two different manifestations. Convergence can successfully bring about a change in trajectory when two diverse elements combine to create a new shared singular trajectory. Alternatively, the convergence of two disciplines' linear vectors that happen to occupy the same space and time momentarily before diverging and continuing their different paths could also be considered a beautiful phenomenon, yet such a result does little to change the individual disciplines. This momentary convergence leaves residual seeds or shrapnel that impact those involved at that time and location. In both cases, the convergence is valuable and made possible by a different convergence of expertise and vision.

In Robert M. Pirsig's book, *Zen and the Art of Motorcycle Maintenance,* there is a passage that carries the often-forgotten subtitle, "An Inquiry into Values," that aligns with our work with clay and robotic fabrication. It is a passage that describes a welder who deploys a high level of craft, material understanding, and skill to repair a chain guard (Pirsig 1975).

I often project the persona of this welder onto ceramicist colleagues I've collaborated within a Harvard ceramics studio over the last 15 years. These experts with tacit material knowledge love to laugh at the statement "Throughout this semester, our group has found the limits of the material and developed a process that pushes beyond those boundaries..." or similarly, that has been spoken by students in every class at every university in which I have explored both clay and robotics. We could not leverage clay-based ceramics without these experts and despite ourselves and our students often ignoring their advice—the utmost respect must be paid to those who have dedicated their life to understanding the material.

At an altogether different production volume, we also must hold reverence for those in industry—both in robotics and clay-based ceramics. Using clay-based ceramic material systems as the framework for understanding robotics we are presented with the opportunity to explore both craft-based production and high-volume production environments. By providing students with an in-depth understanding of material processes in mass production environments, we can start to identify where, how, and to what extent digital intervention is relevant.

In large-volume ceramics production environments we see giant automated factories requiring very low amounts of labor. In a dry pressing environment, many tiles per second are produced and customization in shape is extremely expensive, requiring thousands of units to justify tooling costs. Through industrial extrusion, profile customization is feasible and within reason and extruded linear profiles can be secondarily formed. Slip casting, which is used to create the most complex geometry in all of architecture, the European toilet, seems to be the direct bridge between industrial and craft-based production environments; both environments creating complex geometries through molding. Without robotics, industry applications of ceramics are mostly constrained to either mass-production methods or lower-volume hand building. Design Robotics offers possibilities for greater degrees of variability and creativity.

During the last 15 years, clay has been reintroduced into architectural design pedagogy as a foundation for technology education. Many factors, including financial support from various ceramics industry associations, have made this possible. It is important to highlight that this convergence, of clay, robots, and students is intentional. This convergence creates a framework for understanding the complexities of material systems and the use of variable automated tooling in service to architectural design, building performance, critical thinking, and making. While this effort has had influence on both robotics and clay-based ceramics research and practice, the objective is clear: to provide our students with critical thinking abilities necessary to strategically deploy material and process technology to address real need and to realize exciting opportunities. To this end, we are left with a conflict: what is to be evaluated and how? If the goal is to develop critical thinking in the context of digital material systems, surely the artifact is of less consequence than the process of making. Throughout this compendium of MSD-RAS projects are stories of process told through prototypical tools and ceramic parts. Beyond these objects, a generalizable foundation for practice has developed that enables bold and informed critical thought and action—a practice of understanding more to know less.

## References


Bechthold, M., A. Kane, and N. King. Ceramic Material Systems: In *Architecture and Interior Design*. Basel: Birkhäuser, 2015.

Bechthold, M., and N. King. "Design Robotics." *Rob | Arch 2012*, edited by S. Brell-Çokcan and J. Braumann, 118-130. Vienna: Springer Vienna, 2013.

Robert M. Pirsig. *Zen and the Art of Motorcycle Maintenance*. New York: Bantam, 1975.

# On Learning to Research and Write Scientifically.

— *Billie Faircloth*

There are no robots in this seminar classroom, nor are materials and tools dispersed across workstations. There's no machinic whirring or droning overhead. The scientific research and writing classroom, unlike the robotics lab, is hushed but humming all the same through spells of thinking and talking. In this classroom, RAS students set aside time weekly to be still and uncover the meaning of architectural ceramics experimentation brewing elsewhere in the school. Students bring research doubts and declarations. They test and retest the phrasing of hypotheses, transform bold assertions into study questions, and right-size research implications, wresting truthful statements from hyperbole's temptations. The RAS course Scientific Research and Writing is rooted in an essential, pedagogical premise: as methods to conduct design research evolve, designers must also learn a writing style that makes their work accessible, credible, and reproducible—they must learn the art of scientific writing.

Scientific writing is "a technical form of writing designed to communicate scientific information to other scientists" (Constanza-Robinson et al. n.d., para 1). This definition implies tone and identifies an audience. Interchange the word "scientists" with "design researchers" and imagine how building on existing knowledge serves the design research community and might motivate designers to write scientifically. Architectural theory writing can be argumentative, while architectural history writing can be narrative or descriptive. Scientific writing faithfully documents design experiments, studies, analyses, and their findings. It is fact-based, of the moment, and represents our understanding of the persistently evolving state-of-the-art in architectural technology. RAS faculty maintain that design researchers need research planning and scientific writing literacy to participate in practice/academic/industry coalitions tackling significant planetary challenges.

More than a decade ago, Halina Dunin-Woyseth and Fredrik Nilsson, in their essay "On the emergence of research by design and practice-based research," collected instances where degree programs and architectural practices elevated the design process as a research method, instantiating new academic, professional and transdisciplinary research models (Dunin-Woyseth & Nilsson 2013, 49 - 50). Observing the vital task ahead, Dunin-Woyseth and Nilsson observed, "Because when trying to explain and legitimize in a scientific context the way architectural practice generates knowledge, it becomes clear how immature our field is in relation to more traditional forms of research." (Dunin-Woyseth & Nilsson 2013, 39). The past two decades of researching the materials and methods of assembling architecture differently, computationally, robotically, and through systematic experimentation indicate that architects have become astute knowledge workers. We produce knowledge and write scientifically. Our research platforms are academic-based undergraduate, graduate, and PhD programs, and industry-based firms and labs. As evidence, CumInCAD, the computer-aided design community's subject-based research repository, now has over 12,300 entries, including 9,600 full papers (About n.d., para.1). Originating in

1998, the repository's sustained presence and open access speak to the digital design community's persistence to document and share its knowledge (About n.d., para.2). While it's easy to extrapolate design-studio investigations as forms of research, often the need for study questions, hypotheses, and planned and documented experiments shunts such investigations into pre-research categories. Digital designers have learned to use the scientific method, visualize, diagram, and write down what they've done so that others can do the same. They have also learned to conform their writing into a pattern to share the knowledge they've created, building design research narration into an international movement.

RAS students begin scientific research and writing by engaging previously published research papers to appropriately situate their architectural ceramics research. Researchers refer to this loose and extensive collection of written scientific papers, such as those found in CumInCAD (an index), as the literature. Authored by architects, computational designers, and scientists, published research papers immerse RAS students in current topics on advanced manufacturing, algorithmic and generative design, materials composition, environmental impact, and more. Students begin their spring semester research by reviewing the literature and learning to find, scan, skim, map, and carefully read selected topical papers (Machai & McEvoy 2022, 76). In these papers, researchers methodically document their actions, meaning, implications, limitations, and effuse experimentation and innovation culture. All papers are organized into sections using predictable headings— abstract, introduction, methods, results, discussion, and conclusion—and are often associated with published journals from architectural computation, engineering, robotics, or materials science. Or, they are associated with the international, cooperative meshwork of digital design societies, organizations, and their repositories—ACADIA, DMS, Digitial Futures, CAADRIA, and others— representing a growing and robust design research community.

From this point in the semester forward, students and faculty begin planning and implementing architectural ceramics assemblies

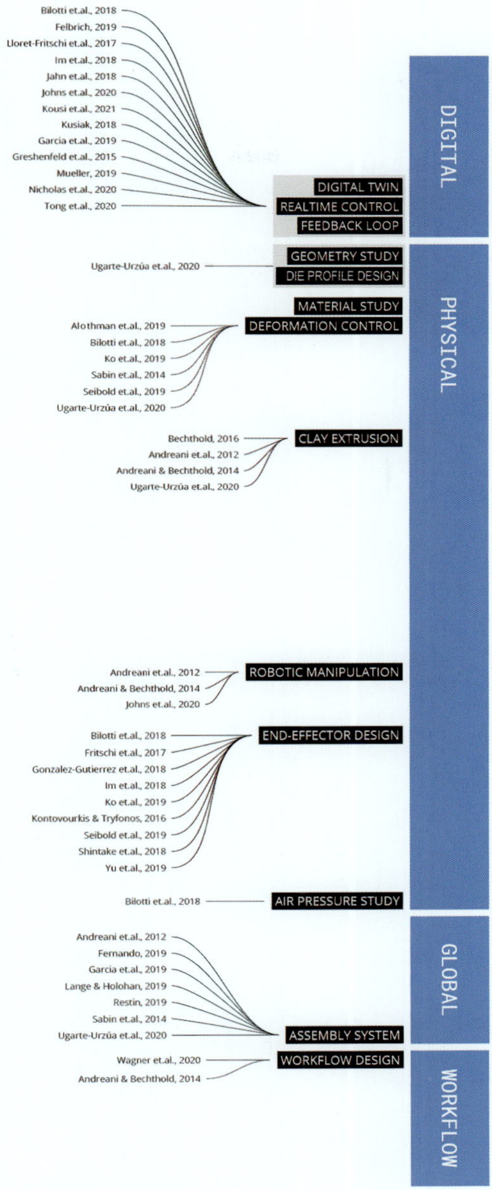


*Student teams learn to conduct a literature review and use literature mapping to visualize the contributions of authors across a multidisciplinary set of fields. This map was created by the Ceramic Forest project team.*

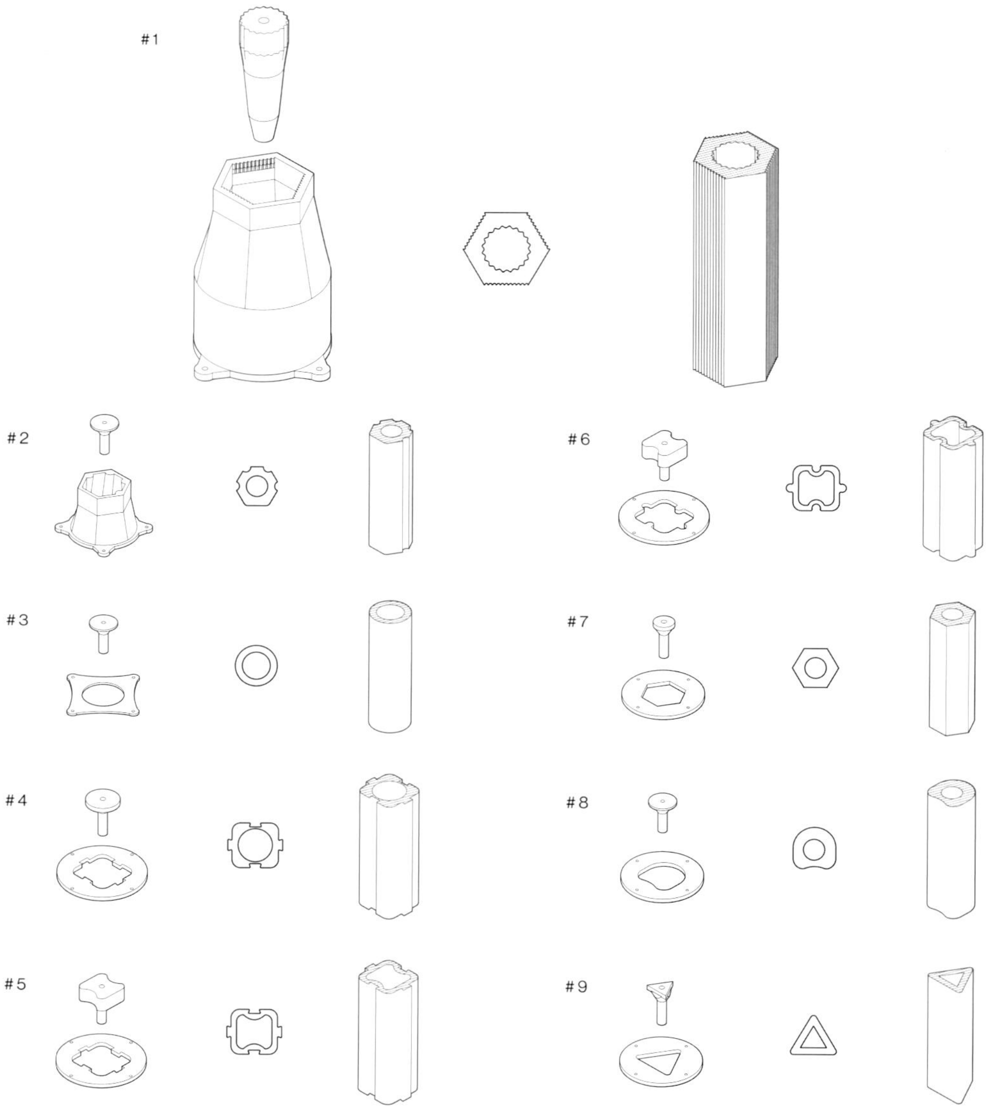


| | #1 | #2 | #3 | #4 | #5 | #6 | #7 | #8 | #9 |
|---|---|---|---|---|---|---|---|---|---|
| **Clay Collapse** | | | | | | | | | |
| **Sectional Area** | *17.7 cm²* | *18.6 cm²* | *15.7 cm²* | *23.9 cm²* | *21.0 cm²* | *20.8 cm²* | *21.1 cm²* | *24.3 cm²* | *16.5cm²* |
| **Air Pressure** | *42psi* | *36psi* | *44psi* | *43psi* | *49psi* | *48psi* | *33psi* | *33psi* | *51psi* |

*Experimental data for different die-cast extrusion nozzle deisgns, documenting the production settings and whether the test was successful or caused a partial or full collapse. This is a simple example of a methodical set of research experiments. By working on a research paper in parallel to project development, research-led design informs such focused sets of experiments. Diagram by spring project team Ceramic Forest.*

research across Material Agencies, Advanced RAS Programming, and Experimental Tooling courses. Scientific Research and Writing supports each student research team and their research project. Through literature synthesis, students also anticipate the value of sharing their research findings within the community.

Overall, students proceed to learn how research is structured, conducted, and communicated; engage critical thinking frameworks; dissect papers section by section; apply research methods; write a full paper for faculty peer review; and talk to practice- and academic-based researchers who learned to research and write about it. Faculty share ambitious goals with the students, which are to:

- Understand the design research process and how it contributes to architectural knowledge production.
- Differentiate between writing types and focus on expository writing's form, purpose, and tone, as demonstrated through primary research and review papers.
- Engage in general scientific writing, emphasizing exposure to the scientific method, research methods, and the parts of a scientific paper.
- Identify a design research topic and project within architectural ceramics.
- Learn to locate and review relevant, published scientific literature and identify current research gaps.
- Learn the components of a research proposal and prepare and present one integrating the full complement of the spring RAS coursework.
- Learn about research and writing ethics.
- Engage a community of researchers and writers in robotics, material systems, and computation.
- Plan, outline, write, and cite a scientific research paper for RAS faculty peer review.

This course upholds the RAS ethos of learning by doing and sometimes, imperfectly, proving that feedback's purpose in design research and writing is often to learn, refine, inspire, and assure quality. Many RAS students begin the course with limited exposure to reading or writing research papers. They are encouraged by the observations of scientist-cum-scientific writer Jari Saramäki, who reveals that even in the education of scientists, there is a shortage of coursework on scientific writing (Saramäki 2018, 1). Several "firsts" will occur over 16 weeks. For some, there's the first scientific paper read; for others, the first literature review, study question formulated and reformulated, hypothesis stated and restated, the first preparation and presentation of a structured research proposal, and the first experiment. There's also the first, second, and third attempt to assign meaning to research where students and faculty discuss project values and goals from the outset (Paul and Elder 2020, 32).

Scientific Research and Writing, its classroom, and the full complement of coursework are an immersive skill-building experience that aims to expose students to new methods, build extensible methodologies, and amplify a designer's capacity to be a credible knowledge worker. It organizes coursework into three consecutive learning streams:

1. ***Students work to prepare a research proposal, presenting it at the first all-course review (weeks one through four).*** *Course instruction begins with a general introduction to the definition, purpose, and components of scientific research and writing. It introduces students to mind-mapping techniques to identify a range of potential subtopics relevant to architectural ceramics and their design, manufacture, and performance. Students learn the importance of locating and reviewing literature by examining several subtopics within the field and aggregating a ceramics research catalog. Week four culminates in teams collaborating to declare a research topic and propose research, including planning for initial investigations, experiments, and prototypes.*
2. ***Students increase confidence in their research topic and ability to plan and conduct research (weeks five through 10).*** *Course instruction continues on research planning. Students apply critical thinking frameworks and establish procedures to collect data, and record observations. Working directly with selected papers, they*

*diagram the structure of individual and interrelated experiments. Students prepare to design and fabricate full-scale ceramic assembly prototypes by presenting a detailed research plan and schedule at the mid-semester review (week 8).*

3. **Students learn how design research is organized and translated into a publishable paper (weeks 11 through 17).** *After instruction on all aspects of paper organization, writing, and publishing, students test Jari Saramäki's writing techniques (Saramäki 2018 6) and engage other guidance to understand requirements for the paper's abstract, introduction, methods, results, and discussion sections. Writing commences with the paper's abstract, state-of-the-art summary, paper outline, and fully cited first draft of a 4000-word research paper contributing knowledge to the digital design research community.*

RAS students are encouraged to wrestle with the research and writing process, engaging leading researchers within this community who authentically and unguardedly share how they learned to read and write scientifically. Design researchers Masoud Akbarzadeh, Laia Mogas-Soldevila, Jenny Sabin, Robert Stuart-Smith, Mette Ramsgaard Thomsen, and others have shared the moment they discovered their research interests, the best writing advice ever received, repeated writing mistakes, and continuing challenges with research and writing tasks.

Writing the perfect paper requires successive drafts to legibly communicate the research aim, specifications, and essential findings. The best-written science papers are akin to a performance, with writing so precise, efficient, and logical that they inspire us to visualize design computation and prototyping choreography and tap out the research steps. Writers saturate these papers with empathy for fellow design researchers but not through florid descriptive language to render research moves visualizable. This technical writing is all-business in staccato rhythms, contrasting with insightful, generous discussion, logically ordered section by section. It builds research confidence because writers understand their audience, vocabulary challenges, and reading comprehension mechanics (Saramäki 2018, 69 - 73).

Scientific Research and Writing does not claim the pursuit of perfect writing. It teaches students to plan and implement research, builds the capacity to write abstracts, outline, and summarize, and culminates in the experience of writing a first and fully cited draft with a research team by the semester's end. This course promotes research and scientific writing as interconnected skills to increase a designer's agency and ability to engage in outcome-driven work. During the first three years, seven out of 12 student teams have revised and published with RAS faculty support through organizations including ACADIA, Digital Futures, SIGraDI, and SimAUD.

- Urban Lung: Exploring The Potential Of Passive Façade-Integrated Vertical Convection Facilitation Systems With Architectural Ceramic In High-Density Urban Environments (Lua et. al. 2024)
- Ceramic Incremental Forming: A Rapid Mold-Less Forming Method of Variable Surfaces (Wang et. al. 2023)
- Ceramic Forest: Robotic Die-Extrusion Variable Forming for Architectural Ceramics (Vakhshouri et. al. 2023)
- Dynamic Slip Casting: An Efficient Robotic Approach to Geometrically Variable Ceramic Part Production (Wu et. al. 2023)
- Light Transmittance Ceramic Design-Computation with Robotics (Dong et. al. 2022)
- Smart Branching: An Experimental Method for Heterogeneous Branching Networks using Non-planar 3D Printed Clay Deposition (Li et. al. 2022)
- Introducing Bespoke Properties to Slip-Cast Elements (Dunaway et. al. 2022)

Several papers document new design-to-production workflows for bespoke ceramic assemblies, the integration of end-of-arm-tooling, data capture, and feedback. Most notably, Scientific Research and Writing helps RAS students learn to access, navigate, and build upon existing knowledge, produce new knowledge systematically and collaboratively, and contribute to and share knowledge within a thriving design research community.

## References


Costanza-Robinson, M., , A. Maxwell, C. Wright, and M.E. Bertollini. "What is Scientific Writing? Write Like a Scientist: A Guide to Scientific Communication." Middlebury University. Accessed June 5, 2024. https://sites.middlebury.edu/middsciwriting/overview/

*About.* (n.d.). CumInCAD. "About." Accessed June 5, 2024. https://papers.cumincad.org/about.html.

Dunin-Woyseth, H. and F. Nilsson. "On the Emergence of Research By Design and Practice-based Research Approaches in Architectural and Urban Design." In *Design Innovation for the Built Environment: Research by Design and the Renovation of Practice*, edited by M. Hensel, 37-52. London: Routledge, 2013.

Machi, L. A. and B.T. McEvoy. *The Literature Review: 6 Steps to Success*. Thousand Oaks: Corwin, 2022.

Paul, R. and L. Elder. *The Miniature Guide to Critical Thinking, Concepts and Tools*. Lanham: Rowman & Littlefield, 2020.

J. Saramäki. *How To Write a Scientific Paper*. Independently published, 2018.


## MSD-RAS Published Papers List


Dunaway, D., D. Rothbart, L. Gwinn, Dr. N. King, and R. Stuart-Smith. "Introducing Bespoke Properties to Slip-Cast Elements: Designing a Process for Robotically Controlled Rotational Casting." In *Hybrids and Haecceities: Proceedings of the 42nd Annual Conference of the Association for Computer Aided Design in Architecture*, edited by M. Akbarzadeh, D. Aviv, H. Jamelle, and R. Stuart-Smith, 244-255. La Vergne: IngramSpark, 2022.

Li, C., Yuan, M., Zilong, H., Faircloth, B., S. Anderson, J., King, Dr. N., & Stuart-Smith, R. (2022). Smart Branching: An Experimental Method for Heterogeneous Branching Networks using Non-planar 3D Printed Clay Deposition. In *Hybrids and Haecceities: Proceedings of the 42nd Annual Conference of the Association for Computer Aided Design in Architecture*, edited by M. Akbarzadeh, D. Aviv, H. Jamelle, and R. Stuart-Smith, 90-97. La Vergne: IngramSpark, 2022.

Lua, X., C. Lia, T. Moa, and Y. Wanga. (Accepted.) "Urban Lung: Exploring the Potential of Passive Façade-Integrated Vertical Convection Facilitation Systems with Architectural Ceramic in High Density Urban Enivornments." In SimAUD (details forthcoming).

Vakhshouri, P., J. Luo, S. Su, H. Tang, B. Wang. B. Faircloth, Dr. N. King, and R. Stuart-Smith. "Ceramic Forest: Robotic Die-Extrusion Variable Forming for Architectural Ceramics." In *ACADIA 2023: Proceedings of the 43rd Annual Conference of the Association of Computer Aided Design in Architecture (ACADIA): Habits of the Anthropocene*, edited by A. Crawford, N. Diniz, R. Beckett, J. Vanucchi, and M. Swackhamer, 381-392. La Vergne: IngramSpark, 2023.

Wang, Y., Y. Liu, R. Studebaker, B. Faircloth, and R. Stuart-Smith. "Ceramic Incremental Forming: A Rapid Moldless Forming Method of Variable Surfaces." In *Hybrid Intelligence: Proceedings for the 4th Conference on Computational Design and Robotic Fabrication (CDRF)*, edited by P.F. Yuan, H. Chai, c. Yan, K. Li, and T. Sun, 499-513. Singapore: Springer Nature Singapore, 2022.

Wu, R., S. Moriuchi, S. Li, A. Chen, J. Anderson, Dr. N. King, and R. Stuart-Smith. "Dynamic Slip Casting: An Efficient Robotic Approach to Geometrically Variable Ceramic Part Production." In *ACADIA 2023: Proceedings of the 43rd Annual Conference of the Association of Computer Aided Design in Architecture (ACADIA): Habits of the Anthropocene*, edited by A. Crawford, N. Diniz, R. Beckett, J. Vanucchi, and M. Swackhamer, 267-277. La Vergne: IngramSpark, 2023.

# INTEGRATING

knowledge through research

*Spring Semester*

## Spring Semester

*The MSD-RAS curricula provides an integrated, holistic approach to learning that does not conceptually distinguish between creative and technical work, instead, the program operates under the assumption that possibilities for innovation are at their best when these activities are considered together. In the final spring semester of the program, students gain knowledge and skills in a wide range of subjects including scientific research and writing, advanced computer programming (for augmented reality, simulation and real-time robot control), material and industrial manufacturing processes, robot tooling, and generative computational design. This knowledge is applied within a single design-research project that is developed by student groups working across all courses. While each course is prepared and run by individual faculty who set a diverse range of assignments, the faculty collaborate to support students in developing a combined project that leverages the knowledge and skills taught in each course.*

*Working in groups, each student's thesis is directed towards the development of a design-research project that can synthesize knowledge gained across classes. Projects are, therefore, multifaceted in approach and arise from student and faculty collaborations throughout the final semester. In addition to regular coordination meetings, course instructors also meet together with students at four moments during the semester to provide collective feedback on thesis proposals and their experimentation, design development, and production activities.*

Courses contributing to this thesis project include:

**802 Material Agencies: Robotics & Design Lab III**
(Design Studio)

Instructor: Robert Stuart-Smith

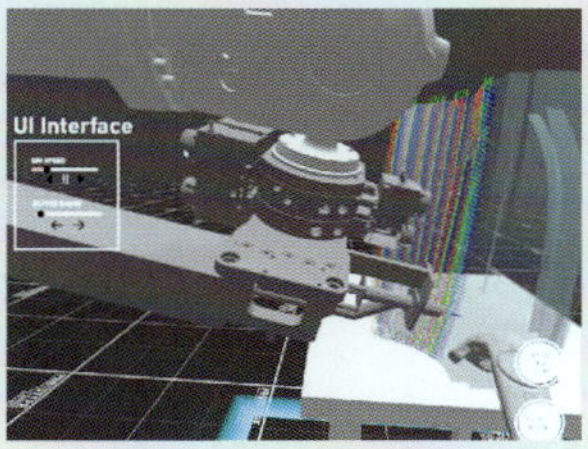


**804 Advanced RAS Programming**
(Real-time Robotics & AR/VR Programming)

Instructor: Jeffrey Anderson
(+ Jose Luis Garcia del Castillo y Lopez in 2020).

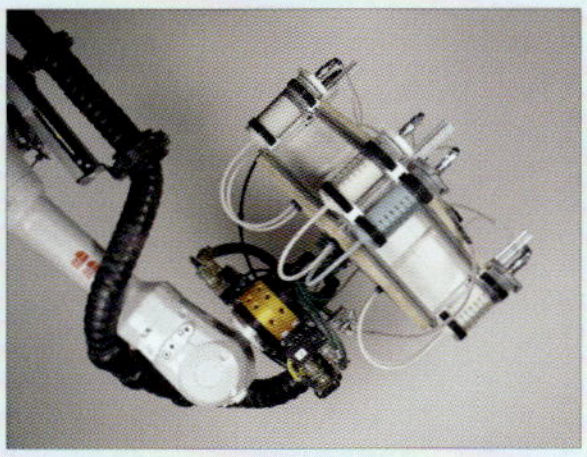

**806 Experimental Tooling**
(Materials/Industrial Processes/tooling)

Instructor: Nathan King

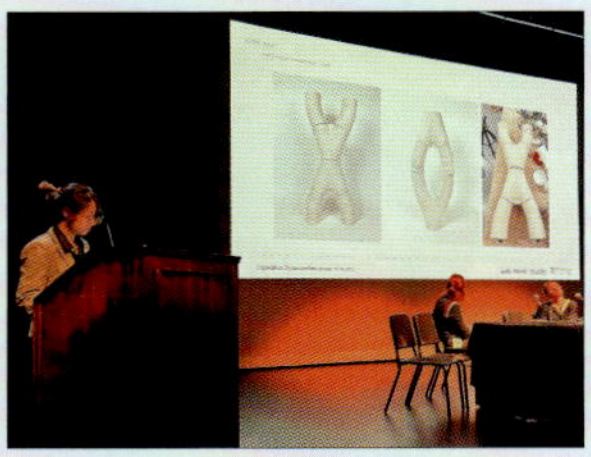

**808 Scientific Research & Writing**
(Methods/Documentation/Writing)

Instructor: Billie Faircloth

# SPRING RESEARCH

Within the spring semester, research-led design projects commence with literature review on design and manufacturing approaches that relate to an open research agenda agreed by the instructors. To ensure these agendas cover significant ground on a topic, they are typically explored across a number of years (up to 3). Literature review activities are followed by a thesis proposal, and a series of design-research experiments that aim to systematically develop a specific and integrated manufacturing and design approach. This knowledge is then be applied to the production of a robotically manufactured, multipart assembled prototype and an accompanying digital design proposal. During the first three years of the program covered in this book, architectural ceramic assemblies were explored as a research focus for architectural façade screen applications (in its fourth year, the program shifted to a different research topic).

### Architectural Ceramics

Spanning across continents and cultures, ceramics has been a key medium in which to explore art and architecture for millennia due to its ability to be easily formed into a wide array of shapes by hand, or through various mechanized forming and casting processes. From Ancient Greek temple friezes and Roman roof tiling to vibrant and colourful ceramics in the work of architects such as Antoni Gaudi or Hector Guimard, ceramics has been realized in diverse functional and aesthetic architectural outcomes. With the increasing use of steel frame construction and the development of skyscrapers in the US in the late nineteenth and early twentieth centuries, terracotta became an excellent cladding material in commercial buildings and a suitable alternative to stone due to its fire resilience, low weight, and versatility in being formed into custom profiles. Louis Sullivan's Guaranty Building in Chicago (1895) is considered to be the first US project to be entirely clad in terracotta. Together with New York's Woolworth Building (1912) by Cas Gilbert, these projects demonstrated ceramics also supported exceptional intricacy and complexity, enabling architectural designs to unify the large scale of a building's volume with the scale of its small ornamental details. Sullivan's work also tested variations in the expression of vertical, horizontal, and tectonic features in a building's overall composition, sometimes highlighting while at other times concealing features such as structural grids, windows, and lintels in relation to overall building-scale design expression.

Sullivan's ornamental features imbued each project with design attributes that are uniquely identifiable. These "haecceities," such as the branching corner of the Guaranty Building's cornice (Van Zanten et al., 2000) or the lintel panels of the Carson Pirie Scott Store (de Wit et al., 1986), are not only distinctive, these artisanal-crafted features brought a smaller, more anthropocentric scale to what were otherwise gigantic buildings that bore no relation to the scale of the pedestrian on the street. As the construction industry adopted more degrees of automation that reduced construction time and cost, bespoke works such as these became comparitively expensive due to their dependence on extensive manual labor, rendering such designs less feasible for use in commercial buildings. Ceramics manufacturing shifted to more automated means of mass production, such as extruded profiles (IRCAM, Renzo Piano), slumped sheets (Cloud9), or as flat tiles (Santa Catarina Market, EMBT). While still able to exploit the variability and vibrancy of ceramic finish and color, mass produced ceramics are mostly constrained to simple, repetitive, and periodically organized geometries and simple forming methods that operate only at one scale. As such, these cladding systems are unable to engage in more integral poly-scalar relationships with the order of a building as a whole, as Sullivan's work explored. Robotic manufacturing offers opportunities to challenge this, by introducing economical, environmentally responsible, and time-efficient means to manufacture the bespoke.

*Students design and prototype their own robot end-of-arm tools (EOAT). Student: Ryan Liao*

A return to Sullivan-era bespoke ceramics allows for more heterogeneous and complex design compositions. In the field of architecture, computation-based design methods can be linked to robotic fabrication approaches to support such design possibilities. Most often such work emphasizes a one-sided, numerically informed approach to manufacturing where computational data directs robotic manufacturing. However, more extensive creative engagement with material and manufacturing concerns is possible. Designers can push beyond geometry and composition considerations to engage in two-way dialogue with material-production processes to enable more diverse forms of design realization and expression.

## Design Scenario: Ceramic Façade

Student groups were tasked with developing designs for a prototypical ceramic façade screen that challenged established thresholds between people, indoor and outdoor environments, and other actors. Speculating on the potential for robotic fabrication to facilitate a more bespoke architecture, teams defined specific design considerations that their façade screen aimed to incorporate (e.g., functional, environmental, material, and ornamental agencies of the design), and attempted to resolve these within a design approach that emerges through the development of a novel means of robotic production.

Each team developed and manufactured a full-scale multi-part assembled prototype together with a larger-scale building façade design that was represented through digital media such as VR/AR models and visual image/drawing representations, animations, and videos. Fabricated prototypes could be between 2'x2'x1' to 6'x6'x2' in scale when assembled, contingent on the specifics of each team's proposed design, production, and MSD-RAS budgetary logistics (students did not have to cover material costs themselves).

A 3D template was shared with all groups that operated as a context-less site, providing a scale and spatial configuration for students to take as a non-binding departure point to work from. Proposals did not respond to a traditional architectural brief or site constraints; instead, designs were developed primarily through research and development of a material manufacturing process and strategies for robot programming and generative/computational design that facilitated the realization of the prototypical design proposal. Each team researched and appropriated an existing ceramic manufacturing method and augmented it through experiment-led research into computational, robotic and material processes in order to speculate and demonstrate new opportunities in façade screen performance and aesthetics. Designs were developed through varying degrees of engagement with the following design-research activities:

- Critical & Creative Approach to Manufacturing: Teams undertook a Literature Review exercise that outlined key precedent research and projects that provided a point of departure for each team to establish a design-research topic that worked with a specific ceramic manufacturing process (such as extrusion, pressing, slumping, additive manufacturing, slip-casting, die-cutting, wheel-throwing/jiggering, sculpting/carving, deforming, glazing, etc). Knowledge taught on these methods in classes informs a series of physical tests that are undertaken to establish possible avenues for augmenting and applying a manufacturing method.
- Computational Design/Simulation: a combination of 3d modeling and computational processes aimed to synthesize multiple design concerns within a 3D design solution and negotiate conflicting design criteria such as solar exposure vs . porosity, surface area vs. volume or structural integrity vs. thinness. Where possible, these methodologies incorporate aspects of the manufacturing process within design from the outset, alleviating the need for industry practices such as design rationalisation or value-engineering (e.g., material simulation or incorporation of manufacturing constraints within the design model).
- Robotic Fabrication: Algorithms were developed to orchestrate robot manufacturing operations that produce variable design effects within automated or semi-autonomous routines to support bespoke production at scale (large-volume production). Methods might incorporate sensor feedback to enable materially sensitive or adaptive behavior.
- Physical dynamics: Material effects caused by robot behavior-based programming, robot tooling, or production knowledge were developed in some projects to convey unique behavioural character and register effects from these contributing technical methods. For example, the motion trajectory, velocity, and tooling of a robot can be utilized to manipulate material in a controlled way, leveraging inherent rheological properties through gravitational and robot motion forces.

Together, these approaches aimed to provide an approach towards a design agency capable of producing novel aesthetic effects while addressing the logistical and environmental efficiencies of bespoke design production.

*Multi-part ceramic assemblage fabricated prototype, developed by students in the ARI Robotics Lab through research followed by a manufacturing production period. Students: Renhu (Franklin) Wu, Shunta Moriuchi, Yinglei (Amber) Chen, and Sihan Li*

## A Marriage of Ethics and Economics

While ceramics is frequently used for the making of small bespoke artworks, undertaking production at the scale of buildings and building component parts poses additional challenges, particularly in terms of time, cost, and waste efficiencies. Fordist production methods met these challenges through repetitive use of dies, molds, jigs, etc., that enabled substantial return from investment only by producing numerous identical parts. Robotic platforms, however, support one-off bespoke manufacturing. While fabricating unique parts with a programmable machine might seem to add almost no additional cost, this is not necessarily true, and great consideration needed to be undertaken in evaluating and developing each project's production workflow. Methods were preferred that generated exciting creative design potentials while also enabling greater efficiency and complexity, and remaining economical, as these could provide more traction in industry.

Participants in the program also went to great lengths to ensure material circularity as much as practical. Throughout these three years of research into ceramic assemblies, clay bodies were repeatedly recycled, re-pugged and re-manufactured. The majority of prototyping was, therefore, a zero-waste enterprise until final demonstration prototypes were produced, which involved firing and glazing clay bodies, changing their long-term material properties.

## Project Outcomes

The final-semester RAS projects presented in the following pages successfully developed novel robotic manufacturing approaches and corresponding design methodologies and applied these in a one-to-one scale prototypical assemblage within the length of a single 16-week semester. This is no small feat, student teams managed to develop a manufacturing workflow and apply it within a short and intensive robotic production timeframe to manufacture a large number of parts for a full-scale prototypical assemblage. Each project's production was executed within an extremely short time frame towards the end of the semester with limited time to learn from manufacturing errors to refine production methods or remake problematic parts. Ceramic production constraints also posed additional scheduling challenges due to the lead time required in production to account for clay bone-drying, bisque and glaze firing compared to other manufacturing processes that are more immediate.

Each team's final manufactured prototype and architectural design outcomes were documented in a project thesis book, accompanied by initial research and experimentation activities. These books aim to capture the incredible depth and diversity of focused work a team of three or four students can achieve across four courses in one semester. In addition to this primarily visual document, each team also developed a research paper that outlined the most novel contribution the project made to architectural research. At the end of the semester, the RAS faculty evaluated these papers in a mock peer-review process and supported strong papers for submission to peer-reviewed conferences or journals. Each team also presented their work to an expert jury with deep knowledge of the semester's research topics before completing their thesis books and research papers, providing invaluable feedback for how these documents could best encapsulate a team's research. Across the three years of this research agenda into ceramic façade screens, all of the research outcomes were of high quality and interest. As is apparent in the work documented in this book, the projects embody both scientific rigor and aesthetic curiosity. As already mentioned in the introduction and elsewhere, several of these projects also had peer-reviewed research papers accepted at world-leading conferences where the work was warmly received. A brief summary of each contributing course's syllabus is provided on the next page followed by a small selection of student work from the spring semester of the program.

## Material Agencies: Robotics & Design Lab II

(Design: Computation & Fabrication

Instructor: Robert Stuart-Smith

Material Agencies questions the role of the designer and their relationship to materials and means of production, exploring varying degrees of design-production agency that allows the designer to engage with the physical world in profoundly different ways. Robotic fabrication can extend production beyond automated routines with degrees of autonomy that enable sensitive material engagement to drive alternative forms of material expression. Robotics and AI technologies also support industrially scalable approaches to bespoke design for manufacture (DfMa) that might address material waste, cost, expand user or site customization, and can respond to a wide range of other design criteria all the while exploring aesthetic opportunities. In Material Agencies, speculative design proposals are developed through fabricated prototypes, and the parallel development of a design-to-production workflow that is explored and demonstrated through computer simulation, modelling and robotic workflows. As such, the course seeks to engage with robotic fabrication and material production as generative contributors to creative design outcomes, whose character is intrinsic to each specific design and production workflow.

During the three years documented in this book, shared across all spring semester courses, students developed a full-scale prototype of a multi-part ceramic assembly for a façade screen together with a larger-scale digital building-scale façade design. Projects were developed outside of a traditional architectural brief or site constraints, operating primarily through research and development of a material manufacturing process, and strategies for robot programming and generative/computational design that facilitated the realization of each prototypical design proposal. These prototypical proposals speculate and demonstrate diverse opportunities related to façade screen fabrication, use, and aesthetics.

## Advanced RAS Programming

(Real-time Robotics & AR/VR Programming)

Instructor: Jeffrey Anderson

Advanced RAS Programming moves beyond the constraints of predefined, offline robotic procedures, to develop frameworks where robot actuation is reactive to materials, environments, and humans, rather than prescripted by them. To afford a greater degree of agency in robotic behavior and human involvement in robotic workflows, students develop real-time control, closed-loop fabrication, and human-machine interaction capabilities using Intel RealSense cameras, Arduino microprocesses, and Virtual and Augmented Reality equipment to envision and explore new workflows for engaging with robotics.

The semester is divided into three parts. In the first two parts, students develop expertise which they then apply to one aspect of their thesis project in the last third of the semester. In the first part, students develop a fabrication routine for a 2D robotic drawing or 3D carving task, which is reactive to visual stimuli using computer vision together with a webcam or RealSense camera-based point cloud data. In the second part of the semester, students utilize VR or AR technologies to create a mixed-reality robot work cell that is calibrated to the robot's coordinate space. This work cell is then used for either a simulation or real-time interaction with the robot to explore human-machine interactions. In the final third of the semester, students develop a custom end-effector and robotic workflow that contributes to the research they are conducting in their thesis.

Over the last three years, students have developed several different forms of human-robot collaborative framework that have enabled humans to bridge technological gaps in automated systems, used computerized systems to augment the human execution of technical tasks, used real-time sensor data to enhance a human designer's decision-making ability, or used 3D scanning technologies to enable a robot's actions to adapt to unpredictable material behavior.

## Experimental Tooling

(Materials/Industrial Processes/Tooling)

Instructor: Nathan King

In architectural practice and during construction, the translation from design to realization is mediated by a range of tools and processes whose development is informed over time by material properties, skill, technology, and culture. Parallel technological developments relating to the way in which things are designed and the way things are made afford new opportunities and challenges for architecture, part customization, user-centered design, and building performance. Within this context, the Experimental Tooling course positions a specific material system as a vehicle for exploring applied research methodologies and an investigation into the opportunities—and challenges—afforded by robotic fabrication techniques. More specifically, Experimental Tooling builds industry-relevant knowledge in robotic and automated methods of production and develops applied research for industrial robot end-of-arm tooling (EOAT) to enhance or augment a material production process and facilitate new design opportunities.

Emphasis is placed on the role of automation and related technologies in low-, mid-, and high-volume production environments. An overview of product manufacturing contextualizes the design exploration afforded by creative applications of robotics along with lectures on product development, production economics, research methods, and other topics. Students are also exposed to industry-standard safety protocols and gain experience interacting with industrial robot hardware, including the internal control systems and input/output typologies, and attain a clear understanding of the delineation between industrial and experimental integration methods. Attention is placed on the design opportunities afforded by new tool development when compared with typical craft and industrial processes. During the course, students design and fabricate industrial robotic EOAT and the programmable controls as part of thesis research. The course includes weekly lectures, discussions, and/or hands-on workshops in the ARI Robotics Laboratory.

## Scientific Research & Writing

(Methods/Documentation/Writing)

Instructor: Billie Faircloth

Scientific Research and Writing helps students build the skills necessary to structure, conduct, and communicate their spring semester research. Students learn to identify a research topic and plan research actions, translating into expository, scientific writing the research they conduct through the courses Material Agencies, Advanced RAS Programming, and Experimental Tooling. They learn the importance of situating their research in state-of-the-art approaches to ceramics and experimentation in robotics, materials systems, and computation. Ultimately, RAS students aim to build credible, replicable architectural knowledge and use writing to connect their work to a thriving design research and practice community.

The course structure is in halves. The first half, Exploring Scientific Research and Writing, teaches the techniques of mind mapping, literature review, and critical thinking so that students can prepare a research topic, hypothesis, and research proposal. The second half, Engaging Writing Methods and Researchers, teaches the scientific writing process and format while students implement their research plan concurrently. The course guides students through research paper components, including the importance of abstracting and outlining, the publishing process, and ethics so that they can complete a full paper suitable for peer review. Professional and academic researchers join the seminar and share their experiences learning to publish, relaying the insights gained through developing their own research and writing practice.

The course's evolution over the first three years of the RAS program emphasizes scientific writing as a fundamental design research skill. It offers an immersive, feedback-rich experience of reading, writing, and editing. RAS research teams have met the course's aim and ambition, successfully planning, documenting, writing, publishing, and presenting their research through several digital design communities, including ACADIA and SimAUD.

# SPRING SEMESTER JURORS

## 2020-2022

### *Guest Jurors:*

Dorit Aviv *Penn WSOD*
Martin Bechtold *Harvard GSD*
Ezio Blasetti *Penn WSOD*
Brandon Clifford *MIT*
Stella Dourtme *SOM*
Shelby Doyle *Iowa State*
Winka Dubbeldam *Penn WSOD*
Jose Luis Garcia del Castillo y Lopez *Harvard GSD*
Laia Mogas Soldevila *Penn WSOD*
Laura Garofalo *CMU*
Rutger Huiberts *KPF*
Jason Kelly Johnson *CCA*
Ryan Luke Johns *Greyshed*
Stefana Parascho *Princeton*
Brian Ringley *Boston Dynamics*
Rhett Russo *RPI*
Jenny Sabin *Cornell*
Michael Sims *ZHA*
Andrew Saunders *Penn WSOD*
Ahmad Tabbakh *Pratt*
Kathy Velnikov *UMich*
Heather Woofter *WashU*
Jenny Wu *Sci-Arc, Oyler Wu*
Masoud Akbarzadeh *Penn WSOD*

# "THESE THINGS SWIRL"

**SLIP-CASTING**

**Davis Dunaway**
**Layton Gwinn**
**Dan Rothbart**

INSTRUCTORS:
Robert Stuart-Smith
Billie Faircloth
Nathan King
Jeffrey Anderson

TA:
David Forero

Slip-casting is an established process for the creation of complex geometries in ceramics and widely used for the manufacturing of sanitaryware and fine porcelain. The phase-changing nature of slip during casting makes the process ideal for the production of complex, standalone components; however, it currently lacks the ability to produce meaningful visual variation between components without the use of an entirely new mold and new slip. This research explores a novel technique for creating bespoke, slip cast artifacts through the use of 6- axis robotic motion to appropriate methods used in plastic rotational molding to slip-casting. By incrementally injecting different ratios of colored slip into the mold while it is rotated, we are able to achieve variable color, pattern, and structure. Because of the highly precise nature of the robotic motion, this variation can be repeated with a relatively high degree of accuracy. In addition, incremental injection of slip allows a full cast to be achieved with a minimal amount of slip, entirely removing the draining process used in traditional slip-casting.

A prototypical façade assemblage was developed that leverages the slip-casting method to produce diverse marbling within a series of 3D components. Before each component is bone-dry, and still malleable, they are assembled into small clusters that bond together when bisque fired, simplifying larger-format assembly post glaze firing. The research aims to introduce new variables into the world of mass-manufactured slip-cast ceramics in the form of controllable color and pattern.

*a*

*Above) several slip-cast parts are bonded and fired as a cluster. Clusters are then assembled together. Opposite) top view of final fabricated protypical assemblage.*

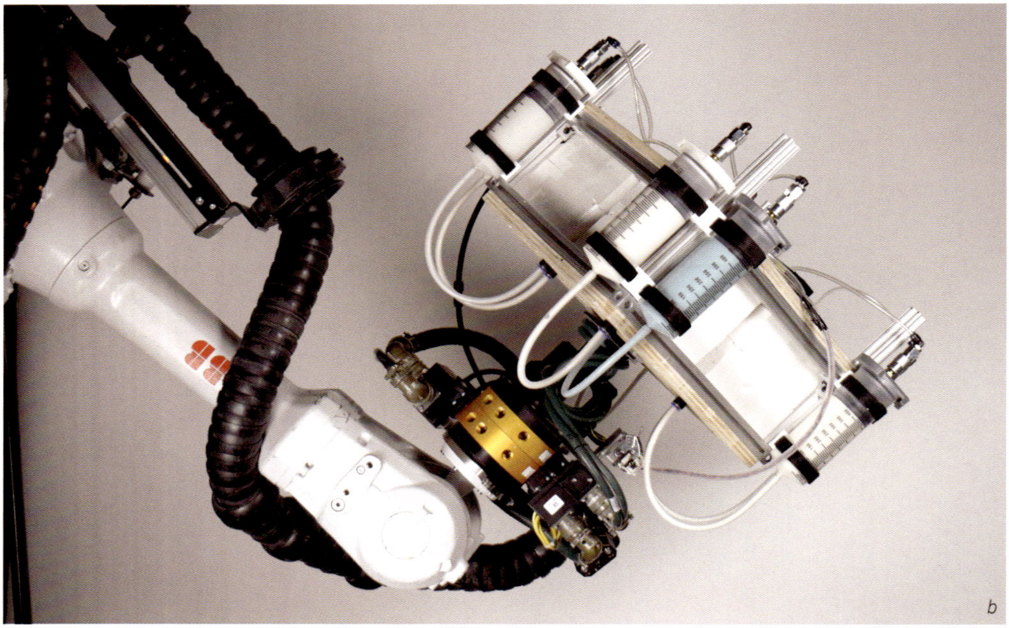

*a) Close-up photo of assembled slip-cast parts showing each one's unique slip marbling and b) custom robotic end-of-arm tool used for roto-molding slip. The tool consists of four slip-cast molds and syringes that inject two slip colors into each mold.*

a

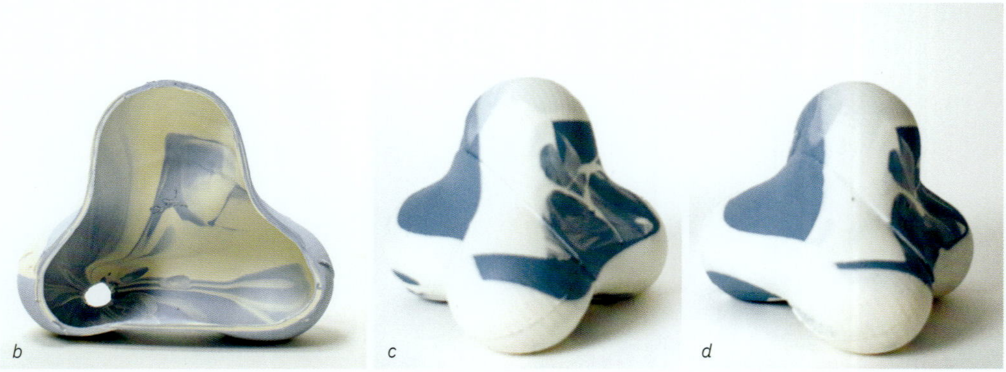

b c d

*a) Photo of final fabricated protypical assemblage, b) a cut through a roto-molded slip-cast part reveals the thin cake (shell) of the part and marbled color, and c,d) two slip-casts from the same roto-molding fabrication action, they are very similar to each other but exhibit slight differences that might relate to their different center of gravity relative to the roto-molding motion of the robot.*

# DYNAMIC SLIP-CASTING

**SLIP-CASTING**

RENHU (FRANKLIN) WU
SHUNTA MORIUCHI
YINGLEI (AMBER) CHEN
SIHAN LI

INSTRUCTORS:
ROBERT STUART-SMITH
BILLIE FAIRCLOTH
NATHAN KING
JEFFREY ANDERSON

TAS:
HADI EL-KEBBI
SOPHIA O'NEILL
MATTHEW WHITE

Slip-cast ceramics can produce unrivalled formal complexity and geometrical precision compared to other ceramic manufacturing approaches. However, its limitation in producing variable components without unique molds presents challenges for the method's integration into architectural practices where variability is often advantageous. This research develops a method to support the production of geometrically varied slip-cast parts from a common mold, through the robotic rotation of the mold during the slip solidification process. Challenges arising from the method, such as slip formation and edge quality, are addressed through the incremental injection of slip during manufacturing on a custom end-effector tool and the development of simulation software to predict slip formation in real-time. The simulation software is also used to generate robot manufacturing instructions in relation to a design input and serves as a design visualisation tool.

The capacity for the approach to produce geometric variation within an automated slip-casting process is demonstrated in a full-scale prototype façade screen assemblage of 28 parts that is comprised of 16 unique geometries that were cast using four different molds. An overall architectural façade screen design of thousands of pieces also uses the same four variations of mold. Altogether, the research demonstrates an approach to the design and manufacture of geometrically varied slip-cast parts for architectural ceramic façade screens that integrates design and manufacturing activities whilst reducing waste and cost associated with the production of unique one-off use molds.

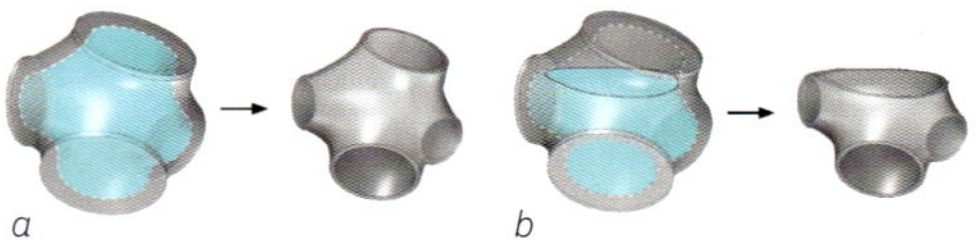

a                b

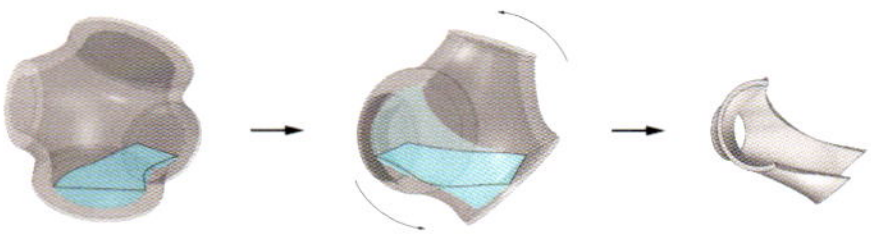

c

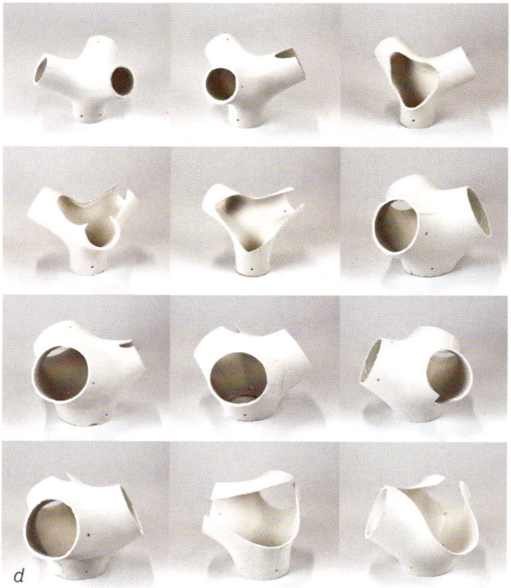

d

*a) Industry slip-cast process requires fully filling a mold with slip, b) partially filled slip mold would create a different "sliced" geometry, c) actual roto-molded slip-cast parts that have a curved boundary and only used a small quantity of slip, and d) robotically fabricated used in final prototypical assemblage. Parts have a variable curved opening generated by robotic rotational-molding.*

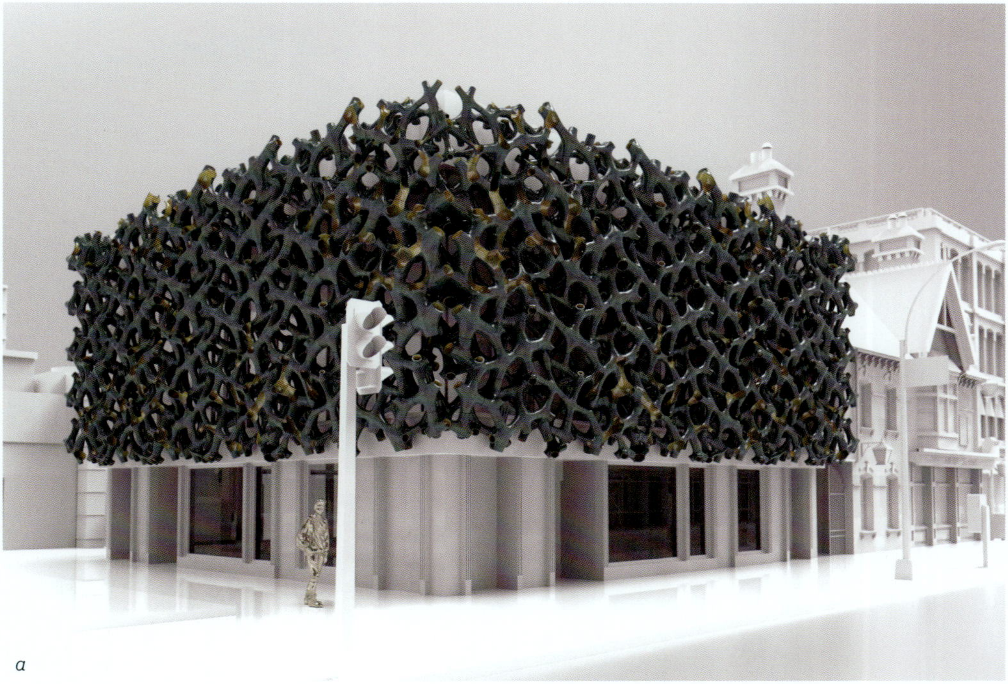

a

b

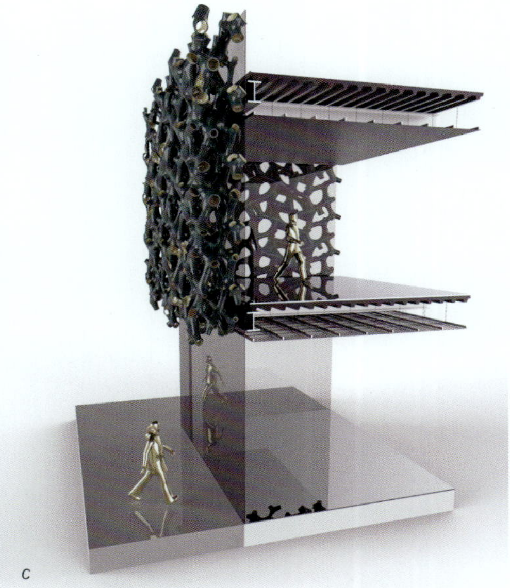

c

*a) Visualization of architectural façade screen, b) robotically fabricated façade screen: multi-part ceramic assemblage prototype, and c) sectional chunk through architectural façade screen.*

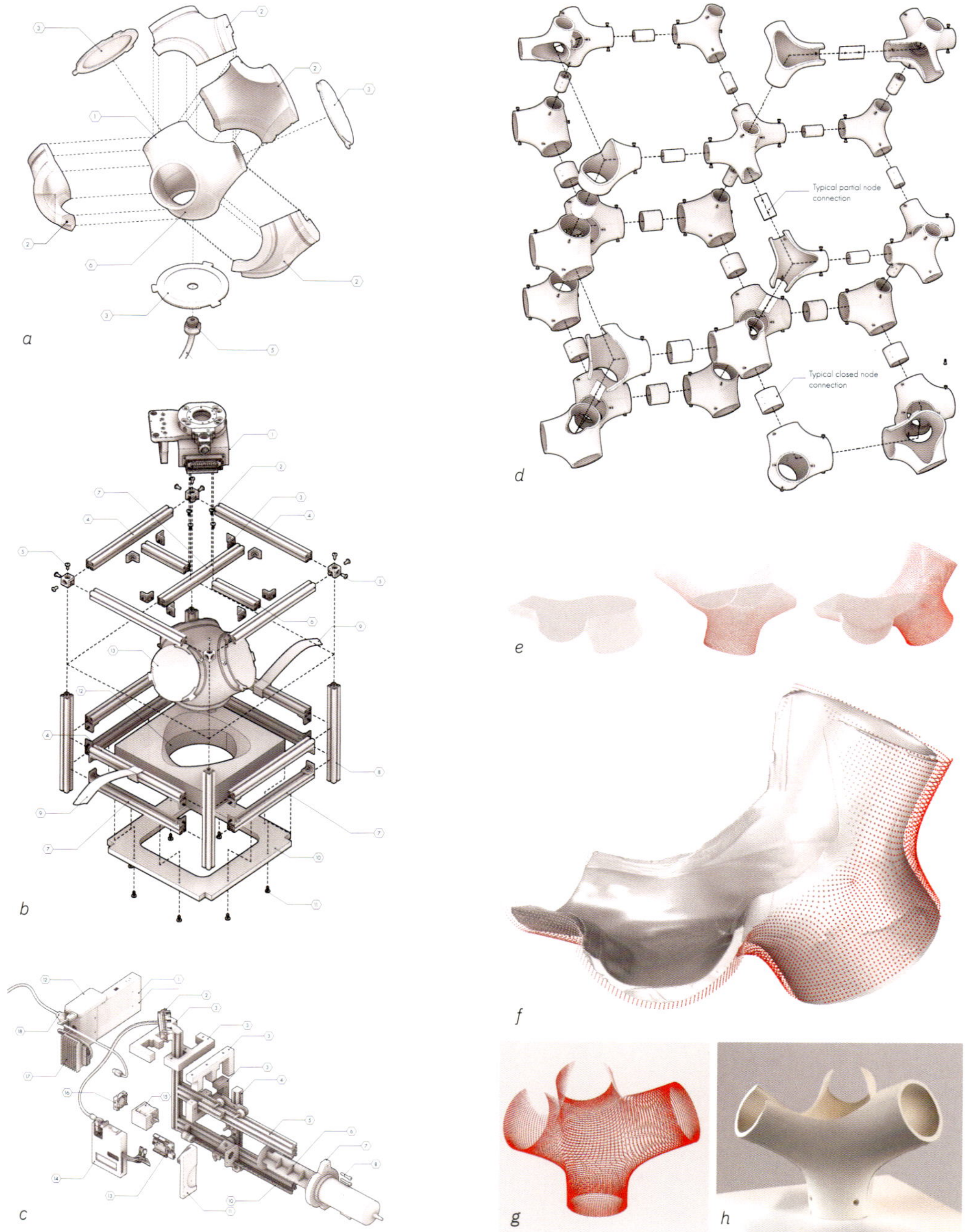


*a) Multi-part slip-cast mold, b)robot EOAT housing slip-cast mold, c) slip micro-dosing hardware attached to robot EOAT that injects small doses of slip for each rotation of the mold, d) exploded taxonomy of 28 parts for fabricated prototype consisting of 16 unique geometries made using 4 different molds, e) three timeframes from custom simulation software, f) validation of simulation model relative to 3D scan of fabricated slip-cast part, comparison of (g) simulation outcome and (h) robotically fabricated part.*

*a) Interior view, demonstrating visual porosity of prototypical architectural façade screen and b) close-up photograph of junctions between several slip-cast parts of varying geometry.*

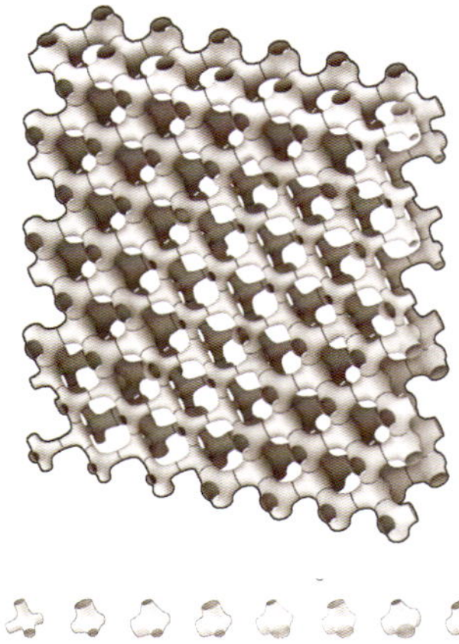

*a*

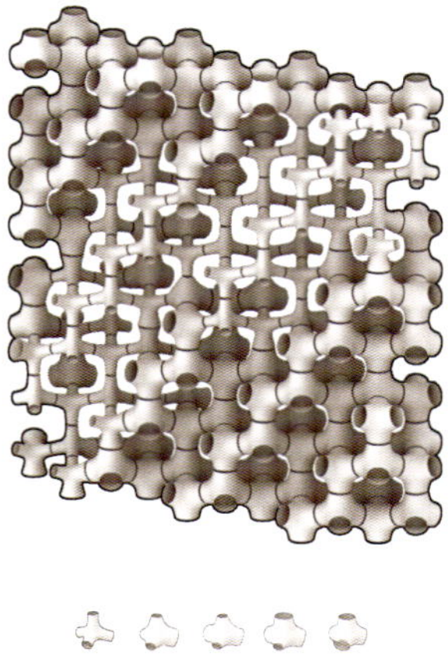

*b*

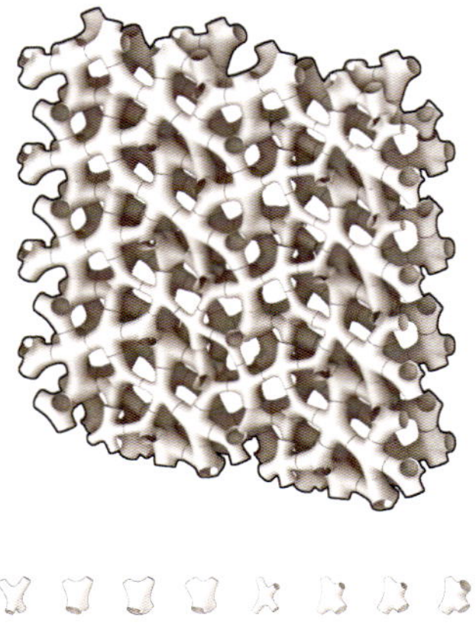

*c*

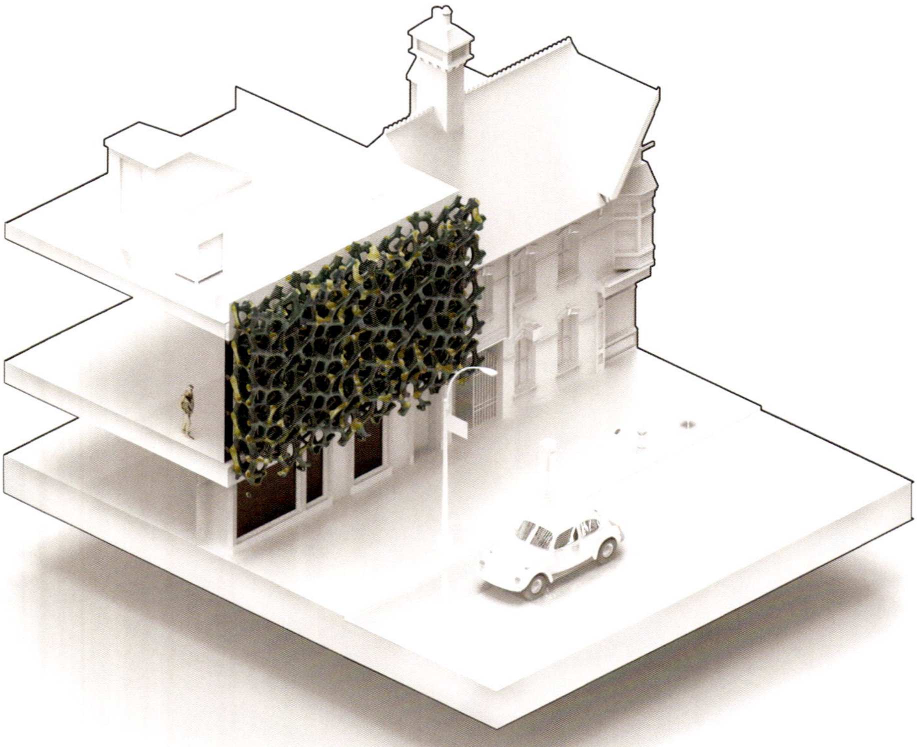

*d*

*a-c) Façade screen studies of different geometrical arrays, each showing an inventory of unique parts which would each require a unique mold and d) aerial view of a sectional chunk from the architectural façade screen.*

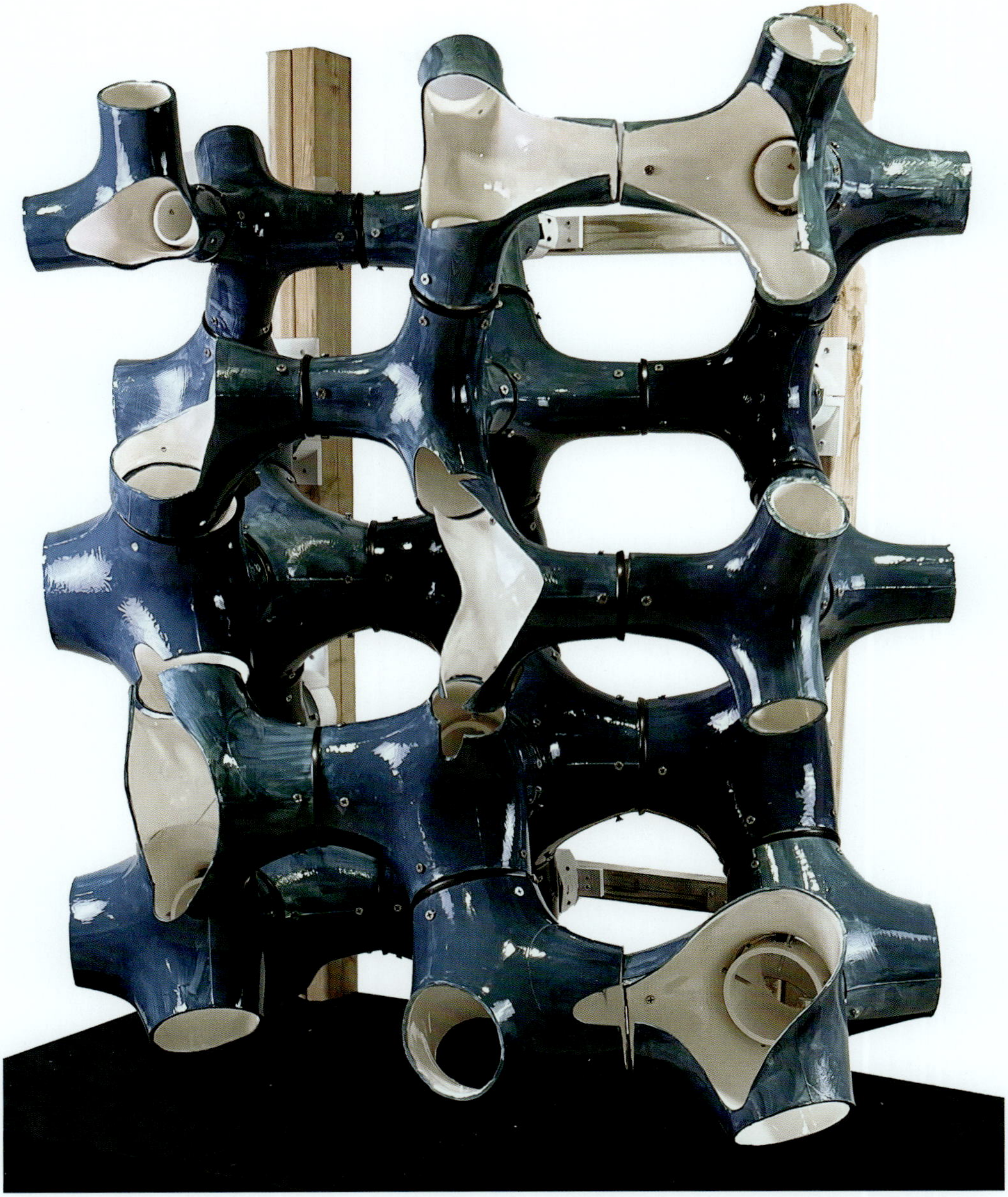

*Robotically fabricated façade screen prototype. This 28-part ceramic assemblage utilized 4 unique molds to fabricate 16 geometrically unique parts.*

# VITA WALL

***SLIP-CASTING + ADDITIVE MANUFACTURING***

*Adhityan Anbumozhi*
*Aisha Alshehri*
*Cem Akgun*
*Mahsa Masalegoo*

*Instructors:*
*Robert Stuart-Smith*
*Billie Faircloth*
*Nathan King*
*Jeffrey Anderson*

*TAs:*
*Hadi El-Kebbi*
*Sophia O'Neill*
*Matthew White*

The field of architectural ceramics dates back millenia. Recently, robotic fabrication methods have broadened the variety of manufacturing techniques available for building applications. Despite advancements in ceramic additive manufacturing and slip-casting, there are limited instances of combining these two distinct fabrication methods. This research explores how different clay shrinkage ratios between bone-dry spherical slip-cast elements and additively manufactured parts can be leveraged to manufacture freely rotatable embedded elements within ceramic façade components, enabling post-manufacture reconfigurability.

An adaptive manufacturing approach was developed that periodically embeds slip-cast elements into an additively manufactured part during its fabrication. The method utilizes computer vision to inform robotic grasping and placement of slip-cast elements without damage to the wet slip-cast part or recently deposited additively manufactured layers. The method demonstrates the potential for the manufacturing approach to be automated to support higher-volume production runs. By combining slip-cast and additive manufacturing methods in a ceramic façade screen, additional functional embedded features become possible, where rotatable hollow slip-cast components could be integrated into a kinetic or repositionable façade system. Such approaches can support the periodic realignment of parts in relation to environmental or species habitat requirements.

The research was applied to a speculative façade screen design that can be reconfigured to support finer-scale adaption to the specifics of a building's urban microclimate than site surveys or environmental modeling can ascertain in advance. A series of rotatable hollow embedded parts are envisaged to provide habitat to various small species as a controlled form of urban re-wilding. This speculative design is also demonstrated in a multi-part façade screen prototype fabricated using the research methods.

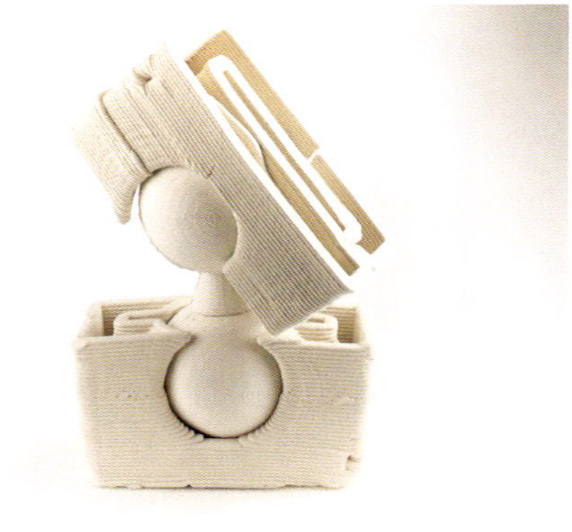

*Additively manufactured (AM) ceramic parts with embedded slip-cast. The AM parts are able to freely rotate around the slip-cast which acts like a ball joint.*

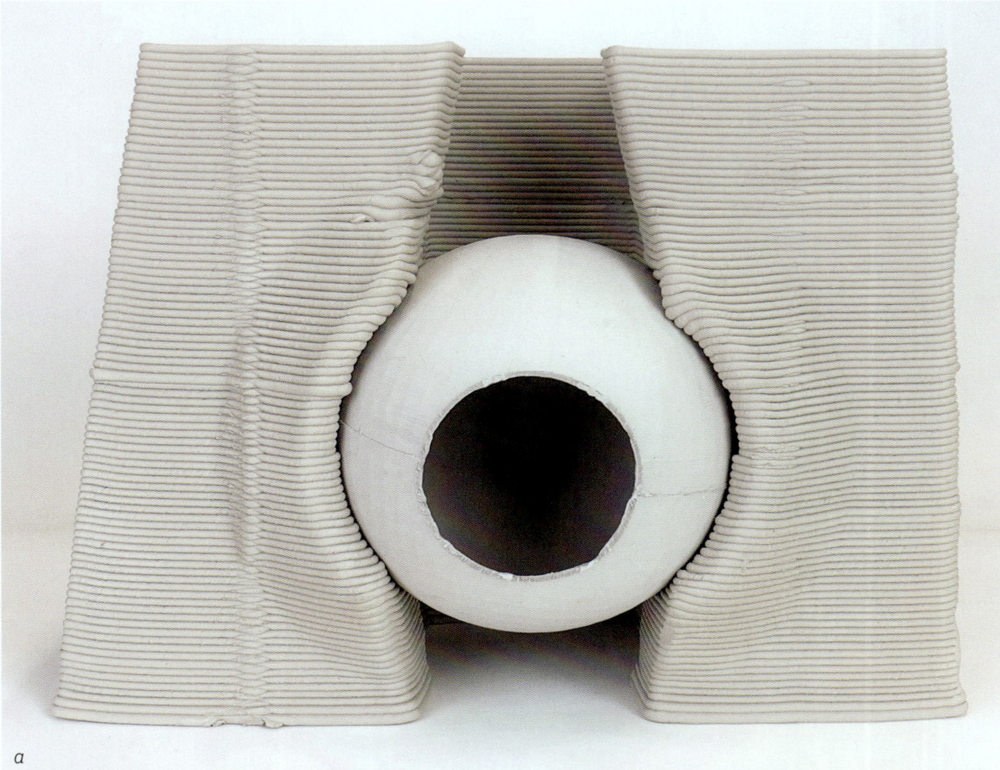

a

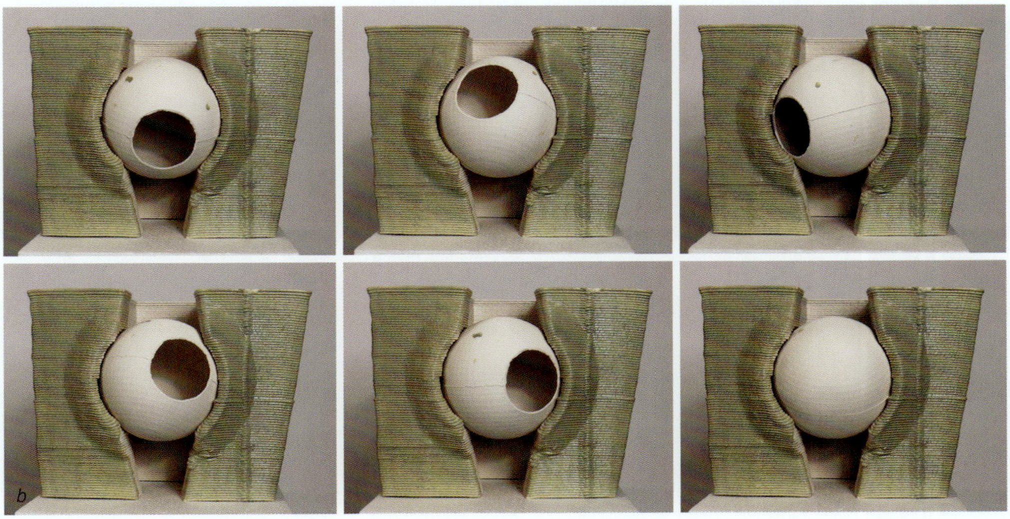

b

*a) Full-scale prototype of a hybrid manufacturing process that embeds slip-cast clay into the additive manufactured (AM) part mid-way through its fabrication. Differential shrinkage in the two elements during bone-drying ensures they stay free from each other during bisque firing, enabling the slip-cast part to rotate freely inside of the AM part. b) The embedded slip-cast part can be rotated to a wide range of*

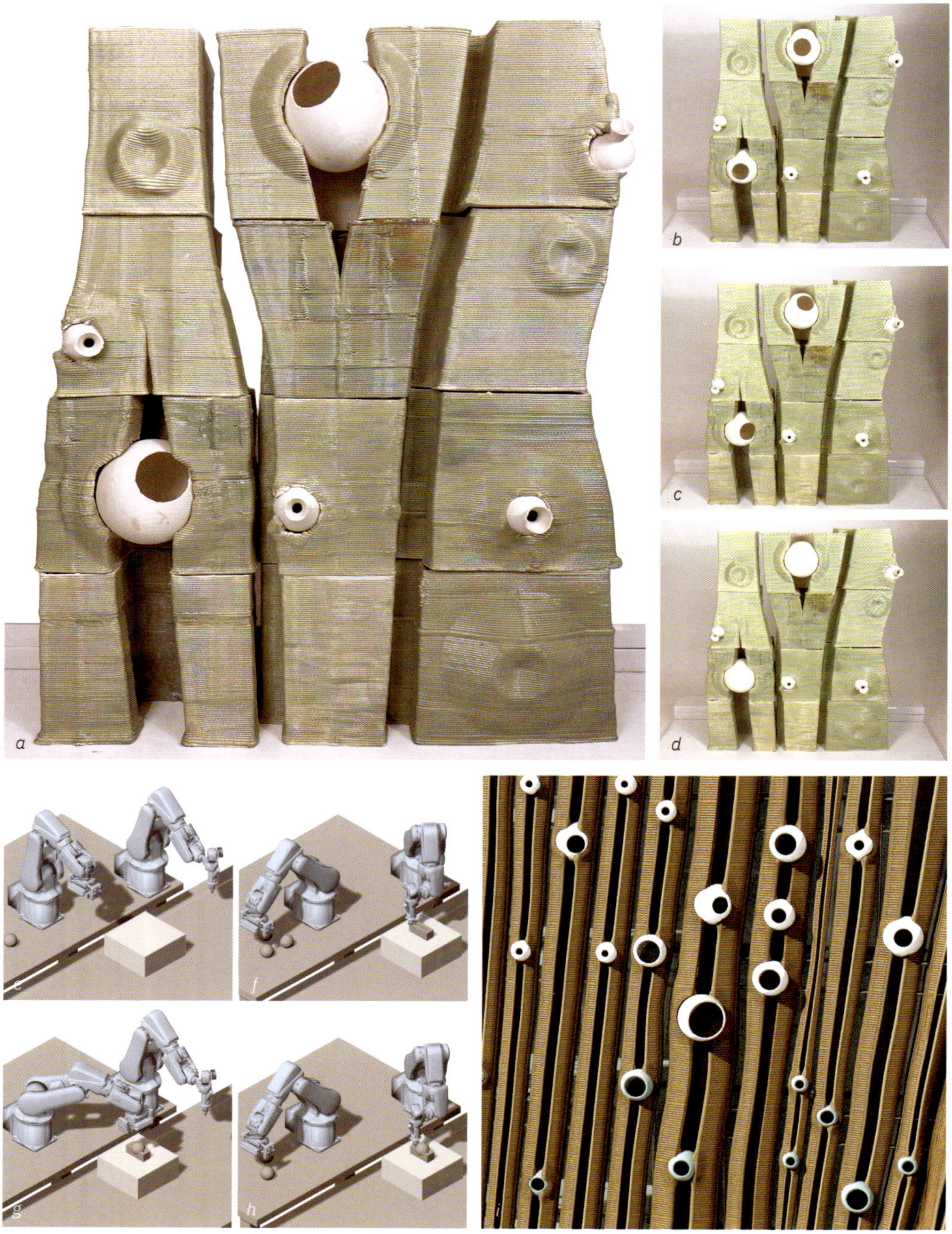


*a) Full-scale multi-part prototypical assemblage, b-d) three different configurations of its embedded slip cast parts, d-g) time-frames from the multi-robot manufacturing framework where one robot does pick and place of slip-cast parts during the other robot's additive manufacturing of clay, and h) close-up render of several clusters of slip-cast parts within a larger facade screen design.*

*An exterior view of a multi-story façade screen design that incorporates a diverse number and orientation of slipcast pockets that might support bird nesting.*

# *HYDROCERAMIC FAÇADE*

***SLIP-CASTING + ADDITIVE MANUFACTURING***

*Christine Khouri Sader*
*Sophia O'Neill*
*Ryan Liao*

*Instructors:*
*Robert Stuart-Smith*
*Billie Faircloth*
*Nathan King*
*Jeffrey Anderson*

*TA:*
*David Forero*

This research explores the embedding of slip-cast geometries inside of robotically deposited clay to create a hybrid manufacturing approach. The method enables a mono-material architectural ceramic façade screen prototype to be developed that utilizes water seepage to enhance passive cooling. The slip cast elements are utilized to create a continuous plumbing network while, additive manufacturing of the prototype's exterior surfaces provides an expedient means of producing geometric variation in relation to building and site related concerns. The research extends clay additive manufacturing capabilities by utilising a robotically deposited sand substrate as a temporary formwork scaffold for the robotic deposition of clay, allowing shallow double-curved surfaces to be fabricated that would otherwise collapse without support, and alleviating the need for wasteful formwork methods as the sand involved is able to be reused.

Clay slip-casts are embedded between two layers of robotically deposited clay. A robot control method and tooling was developed for the deposition of the sand mold, and for 3D scanning and conformal clay deposition to adhere to both sand surfaces and embedded slip-casts. Bisque fired prototypes were filled with room temperature water and tested for their cooling capacity as they seeped the water to their exterior surfaces. Results demonstrated that the seepage's exposure to air produced a substantial cooling effect. A select number of slip-cast molds were designed to support a degree of variation when used in various assembled combinations. The robotically deposited clay extends this variation by supporting unique, non-standard surface geometries around each slip-cast part. To explore a larger speculative prototypical façade design, a computational method was developed to control the distribution and orientation of the slip-casts in relation to solar radiation exposure and prevailing winds. A water seeping façade screen design is proposed that arises from the custom-developed manufacturing method that was informed by, and hybridizes ancient pottery practices with robotic additive manufacturing.

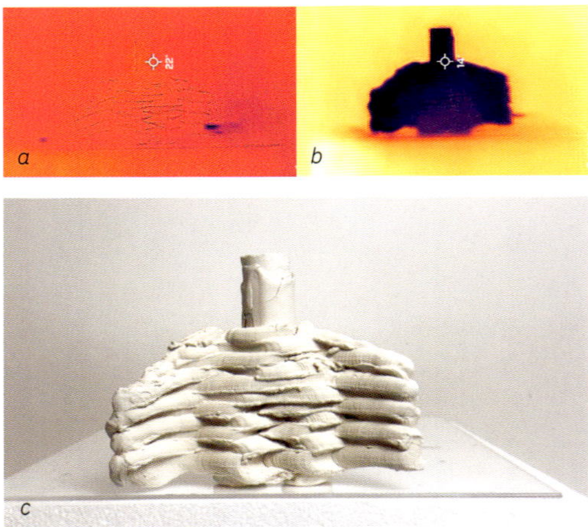

*a-b) Thermal imaging of water seaping from hybrid slip-cast and 3D printed component demonstrating a reduction in temperature of eight degrees centigrade, c) Hybrid slip-cast and conformally 3D printed ceramic facade screen test component,*

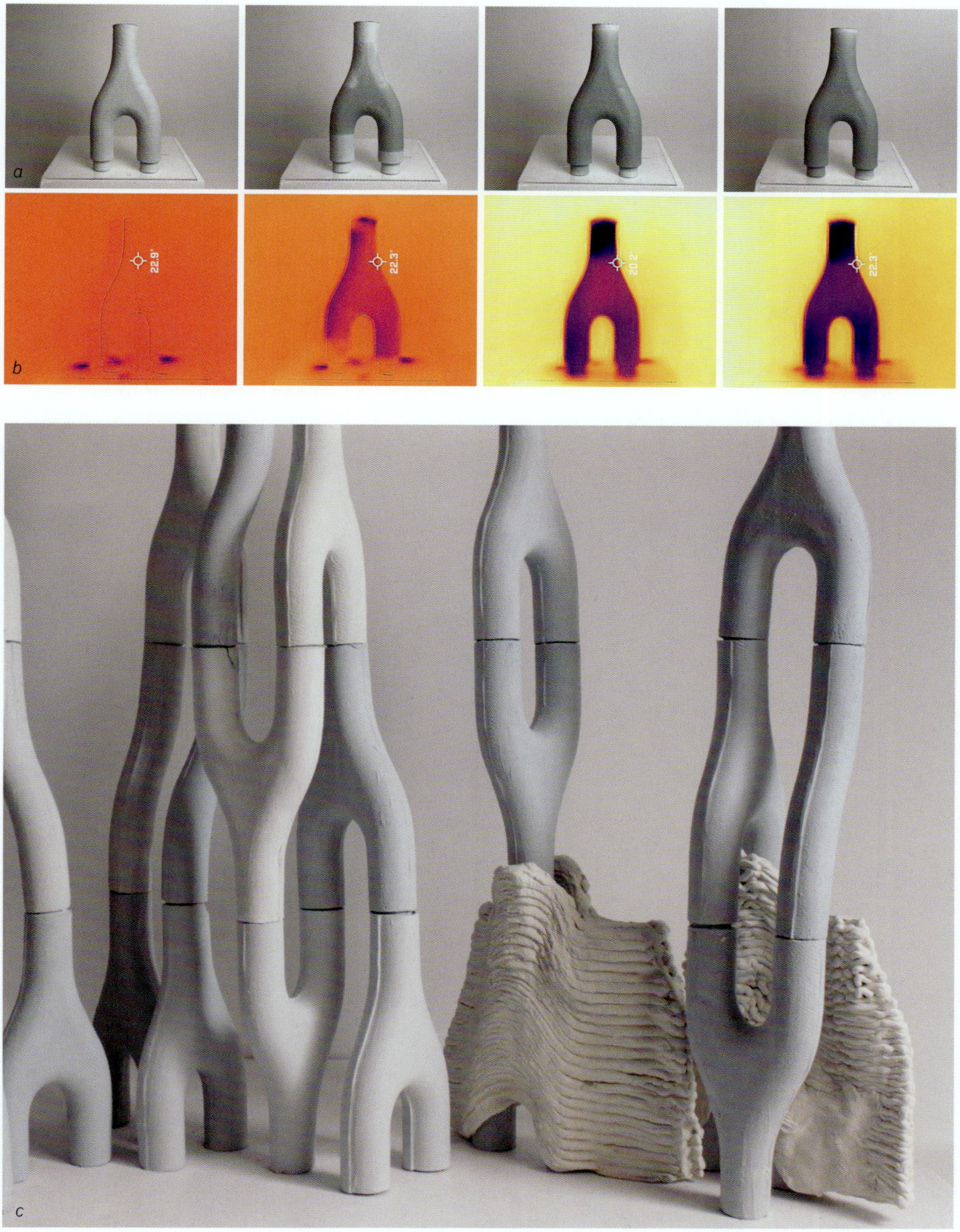


*a) L to R: time sequence of internal water supply seeping through a slip-cast façade screen component and evaporating, b) L to R: thermal imaging of each time-frame reviews an effective evaporative cooling process that drops well below room temperature, and c) photo of multi-part assembled prototype showing conformal 3D printed ceramic parts covering a region of slip-cast elements.*

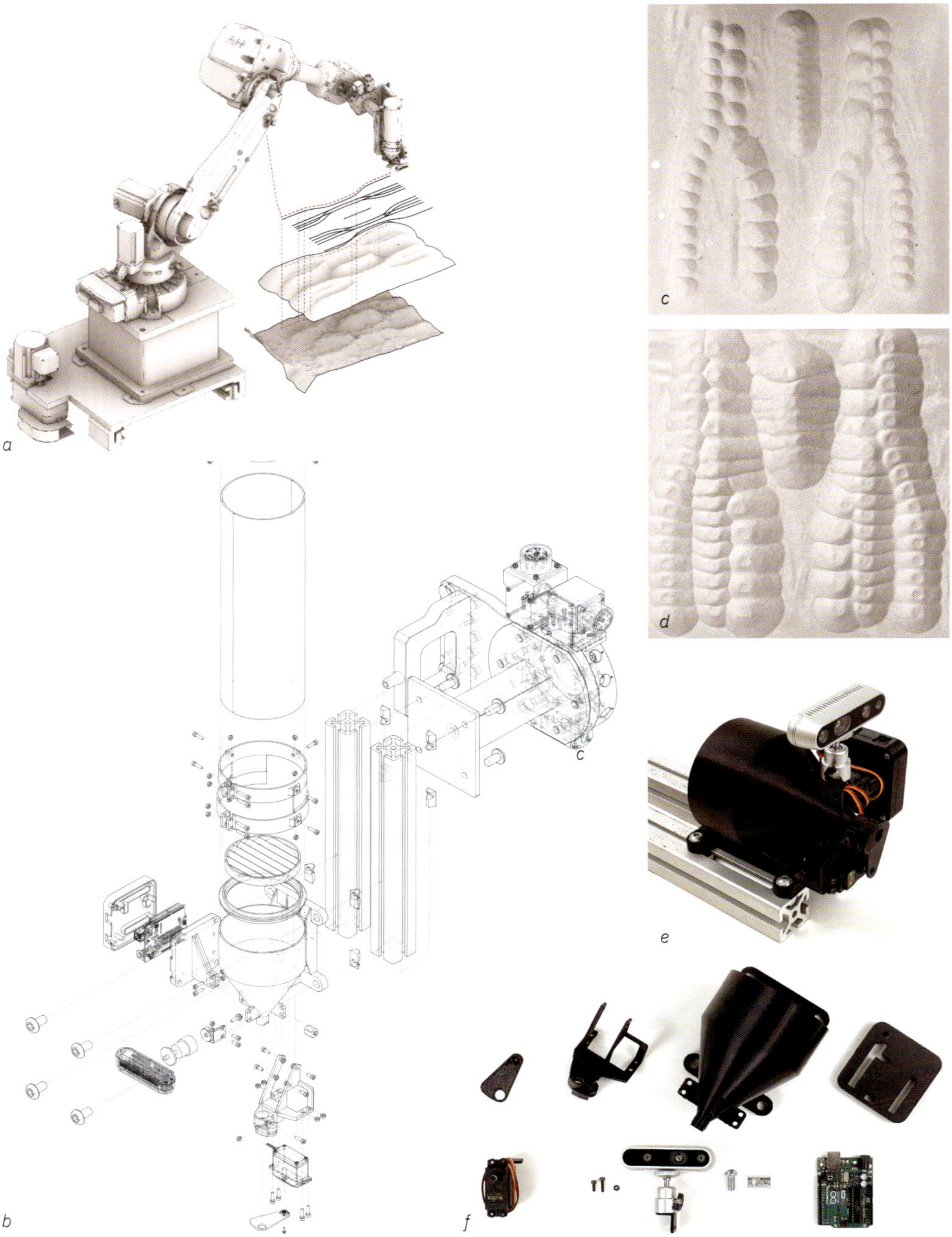


*a) Robotic fabrication method for waste-free, reconfigurable sand-based formwork using a custom end-effector tool that controls sand deposition, b) exploded drawing of EOAT design, c,d) reconfigurable sand-mold, e) assembled EOAT, and f) exploded kit of parts used in a custom robot end-of-arm-tool (EOAT) with sand deposition and 3D scanning capabilities,*

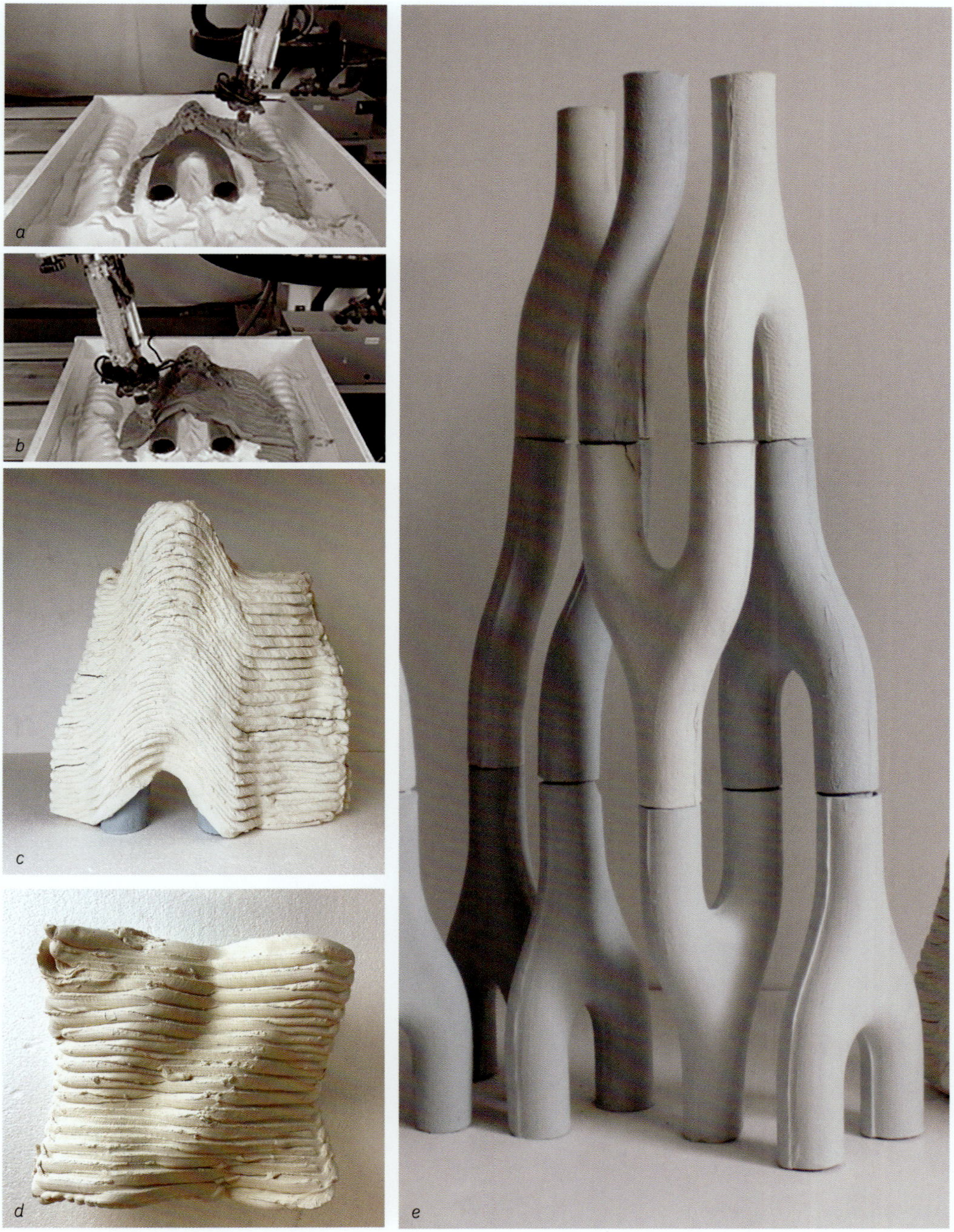


*a,b) Conformal 3D printing of clay ontop of slip-cast component and sand formwork, c) hybrid ceramic part comprised of conformally printed and slip-cast clay, d) conformal printed ceramic wall part, and e) close-up of slip-cast façade screen components.*

# NON-PLANAR BRANCHING

*ADDITIVE MANUFACTURING*

*CICI CHENXIAO LI*
*ZILONG HAN*
*(JUNE) MINGYANG YUAN*

*INSTRUCTORS:*
*ROBERT STUART-SMITH*
*BILLIE FAIRCLOTH*
*NATHAN KING*
*JEFFREY ANDERSON*

*TA:*
*DAVID FORERO*

Clay extrusion 3D printing offers opportunities for innovation in the architectural ceramics industry which currently relies on manufacturing methods that are practically and economically constrained to large-volume production runs. However, due to clay's plasticity and semi-liquid properties, technical challenges remain in finding printable geometries, scaling components, and in developing methodologies for multi-robot cooperation. This research explores the customization potential of clay extrusion 3D printing by adopting a non-planar deposition method for branching structures composed of cylindrical components. It focuses on workflows for printing truncated and bifurcated branches, exploring the relationship between branching component equilibrium and non-planar deposition parameters. These experiments focus on fully controlling the non-planar deposition method while taking advantage of the automatic control granted by a 6-axis industrial robot.

The research is demonstrated in a 1.3m high ceramic assemblage prototype and a multi-layered building façade screen design, where the robotic manufacturing method is leveraged to create a highly bespoke, variable façade through an approach that holds potential for such large-scale industrial applications.

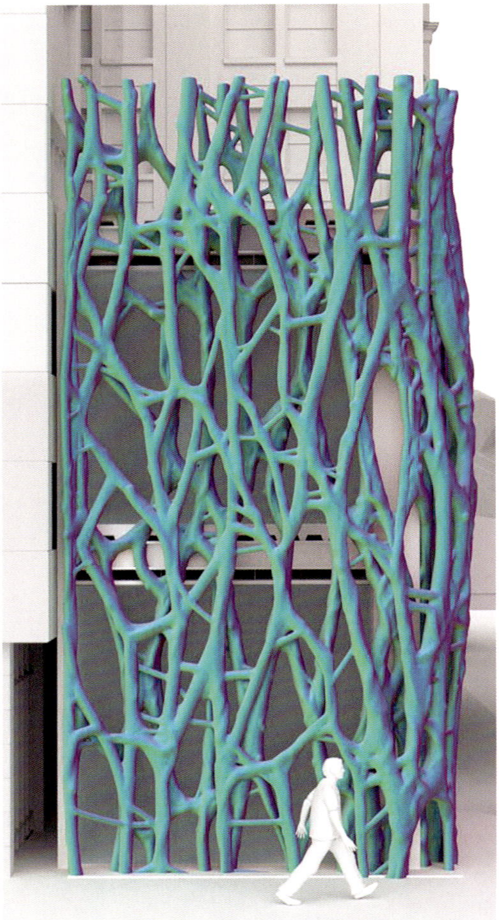

*Elevation of multi-layered additively manufactured ceramic façade screen.*

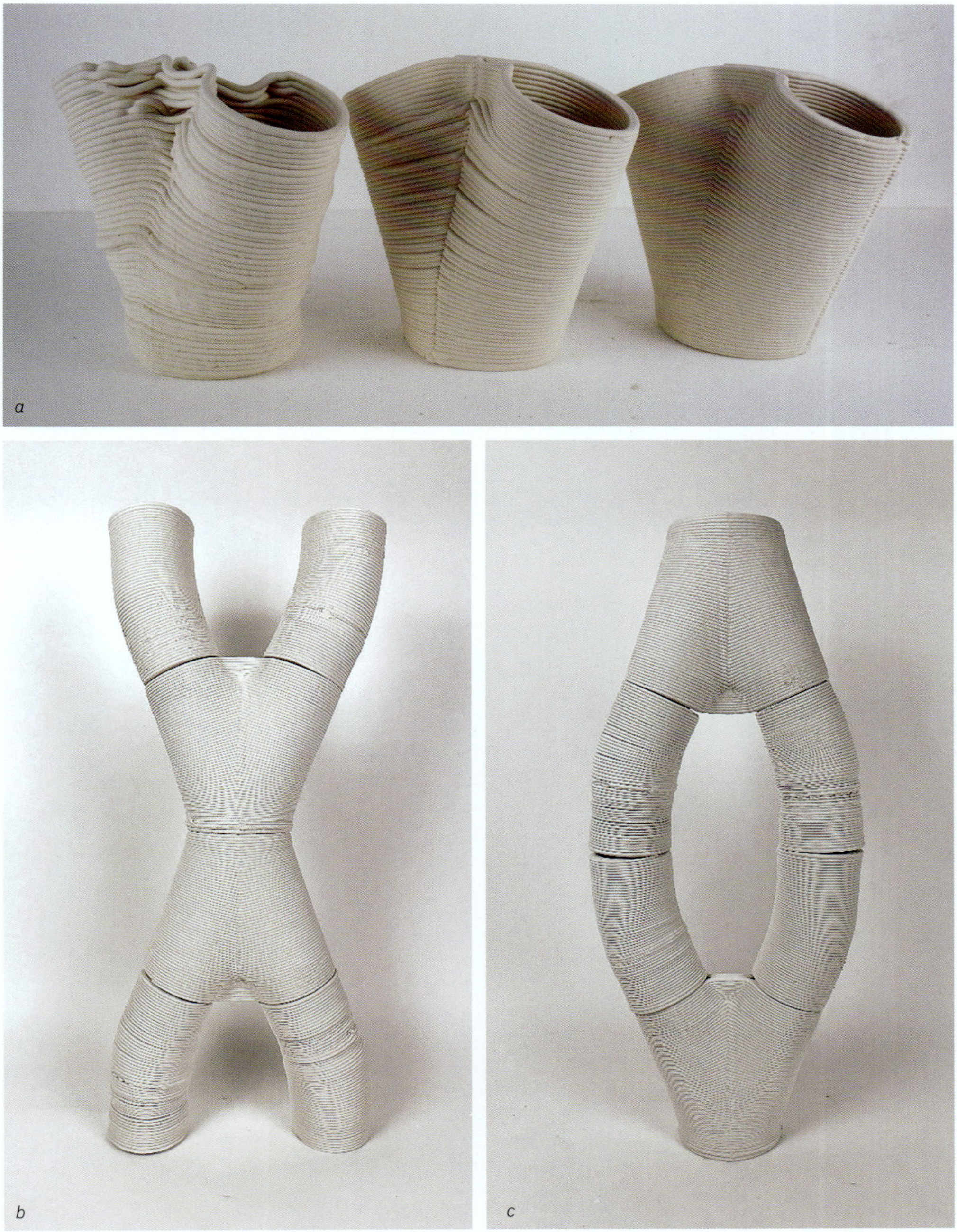

*a) Additive manufacturing of inclined hollow branching elements in wet clay is challenging. The research developed substantial improvements in technical approaches to manufacturing such parts as shown from L to R, early studies through to final parts, and b,c) two multi-part assembled prototypes testing both manufacturing tolerances and the ability of the parts to be post-tensioned together.*

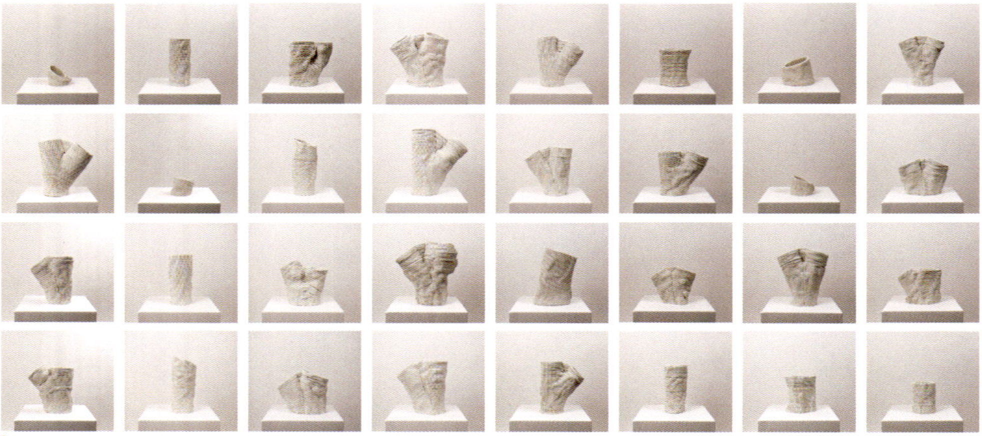

a

b

a) Final family of individually unique fabricated elements, most of which are branching, and b) final assembled ceramic façade screen prototype with an additional algorithmic approach to surface texture and depth being explored.

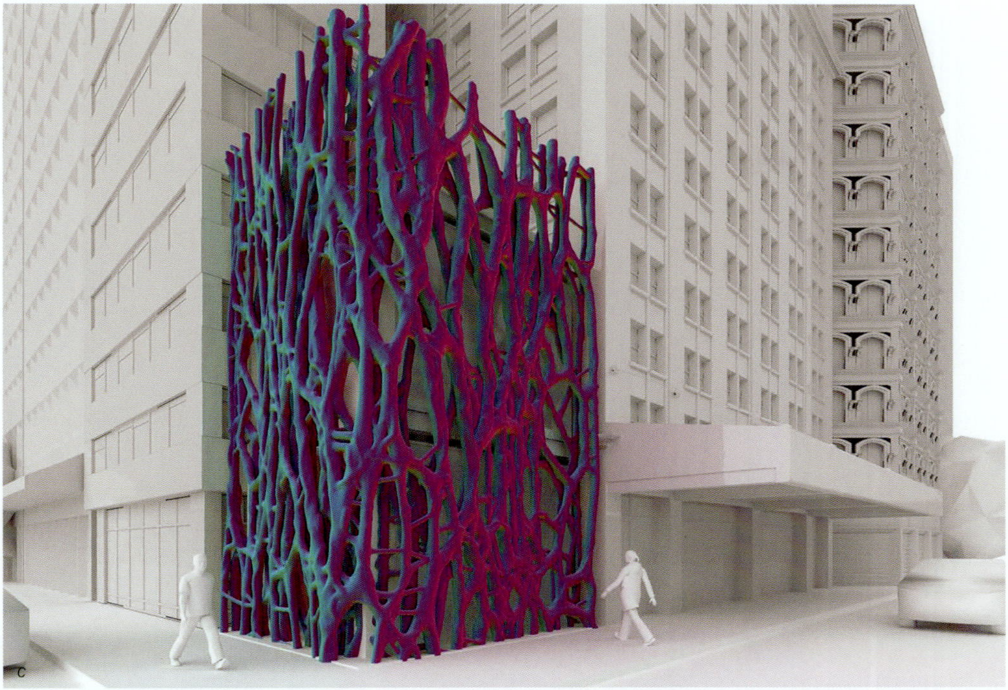

*a) Exterior view of architectural façade screen and b) interior view of architectural façade screen.*

# DISTORTION

***ADDITIVE MANUFACTURING***

***Claire Moriaty***
***Deon Kim***
***Geng Liu***

*INSTRUCTORS:*
*Robert Stuart-Smith*
*Billie Faircloth*
*Nathan King*
*Jeffrey Anderson*
*Jose-Luis García del Castillo y López*

*TA:*
*Patrick Danahy*

Distortion explores dynamics in both fabrication and design activation. Industry forming methods for architectural ceramics operate through industrial means of mass production that cannot achieve bespoke architectural production without wasteful and costly practices. While additive manufactured ceramics can be utilized for variable part fabrication there are geometries that are not easy to fabricate in wet clay, such as shallow arching curved surfaces which are prone to collapse without a mold for support. This research develops a method for additive manufacturing using reusable fabric form-work to enable shallow-profiled geometries to be custom-fabricated in ceramics.

Due to the complexity of the dynamics operating between the extrusion of wet clay and its impact on the form of the elastic fabric, an initial method involved additively manufacturing flat in 2D ontop of the fabric before slumping the manufactured outcome. This method was then augmented to enable 3D manufacturing through the development of a real-time robot manufacturing process that could adapt the position of a robot extrusion point to gravitation-induced changes in the form-work profile caused from the settlement of already extruded clay. Both methods demonstrated qualitative visual character that was intrinsic to the manufacturing approach.

As is evidenced in the prototypical panels, there is an artisinal-like quality produced from a degree of inderterminancy in the material-physical forces produced in the dynamic interactions of the wet clay body, elastic fabric together with the motion trajectory and velocity of the robot. Variable applications of glazing were explored as a means to highten these imperfections, producing a unique form of material expression in the work.

The research method was also explored in a speculative proposal for a subway feature wall system that was envisaged to be capable of producing acoustical effects caused from sound reverberation behind openings in the panels after exposure to high-velocity air movements induced from the passing of trains in the enclosed space of the subway. In this sense, the project embraces dynamics in its manufacture and use.

*a*

*a) Partial chunk of an additively manufactured panel revealing the variable depth in its sectional profile.*

*a,b,c) Additively manufactured ceramic wall panels, b,c) close-ups of panels post bisque and glaze firing.*

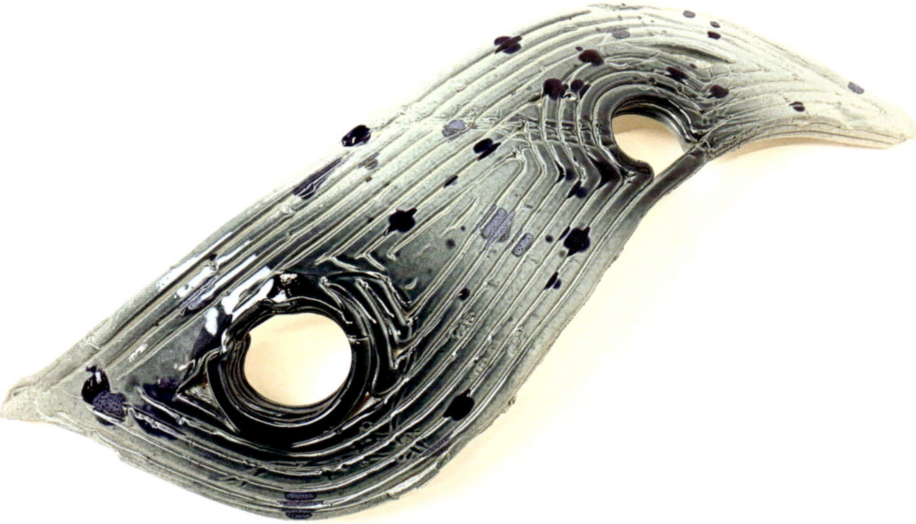

a

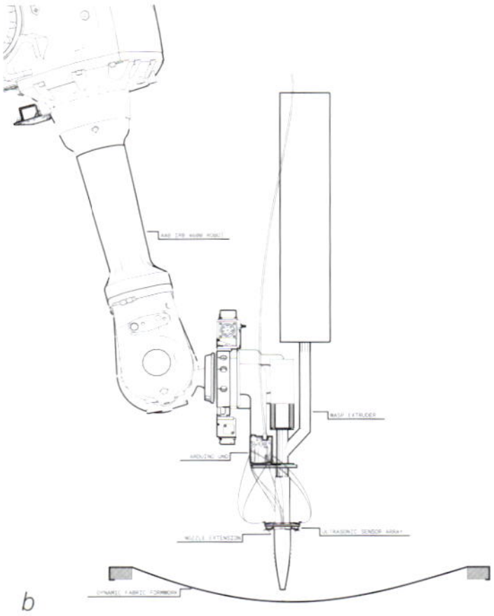

b

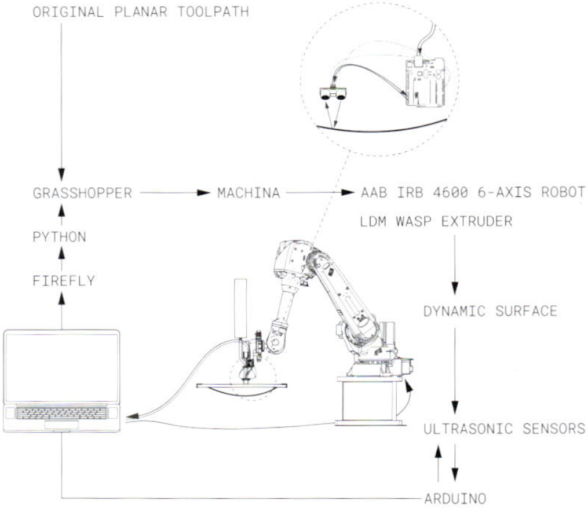


c

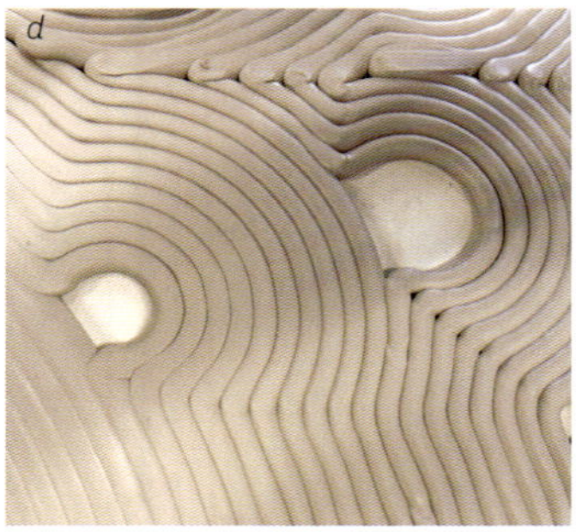

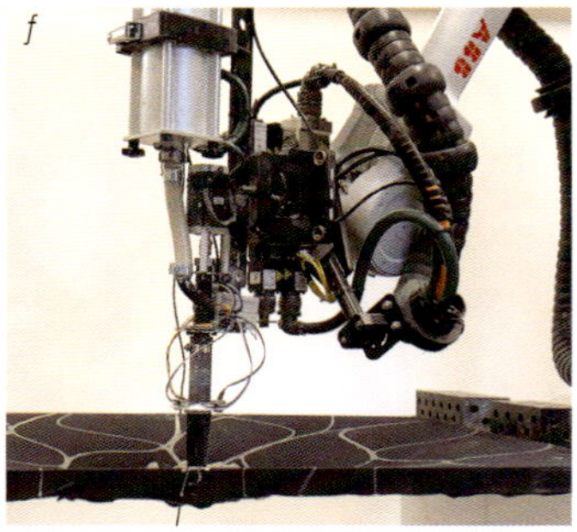

*a) Fabricated prototype wall panel, d) 2D manufactured panel prior to slumping, and b,c,e,f) development of conformal additive manufacturing capabilities where sensor-feedback enables a robot extruder tool to adjust its height in relation to slumping caused by depositing clay on a suspended elastic lycra surface.*

*a) Prototypical ceramic wall panels, b) material-physics computer simulation of funicular form-finding of additively manufactured parts, c) 3D chunk of assembled ceramic wall system, and d) interior view of a speculative feature wall in subway system.*

# ROBOTIC PROMETHEUS

***ADDITIVE MANUFACTURING***

*Grey Wartinger*
*Matthew White*
*Jiansong Yuan*

*INSTRUCTORS:*
*Robert Stuart-Smith*
*Billie Faircloth*
*Nathan King*
*Jeffrey Anderson*
*Jose-Luis García del Castillo y López*

*TAS:*
*Patrick Danahy*

Urban development is responsible for some of the greatest local and regional extinction events, while also creating long-term threats of habitat loss for those species that do survive. Building façades pose a particular risk to bird populations. Bird strikes – collisions with reflective glazing—and loss of fly-through perching habitat in urban centers greatly impacts migratory species, many of which travel directly through prominent cities such as Philadelphia on their journeys. Robotic Prometheus explores the notion of a multi-species façade screen, by creating habitable spaces for bird nesting and perching while providing human privacy and daylighting.

A design and production approach to architectural ceramics was developed that goes beyond established industry methods of mass-production that are predominantly constrained by extrusion dies and molds. The research develops a 6-axis additive and manipulative ceramic manufacturing approach that is capable of producing a heterogeneous façade that embodies diverse surface texture, nesting locations, and ornament. Robotic Prometheus engages with the plasticity of clay through a behavior-based robotic motion approach that leverages agent-based programming and custom robot end-effector tooling to realize a bespoke architectural façade screen design that is inseperable from its specific robotic fabrication and material production processes.

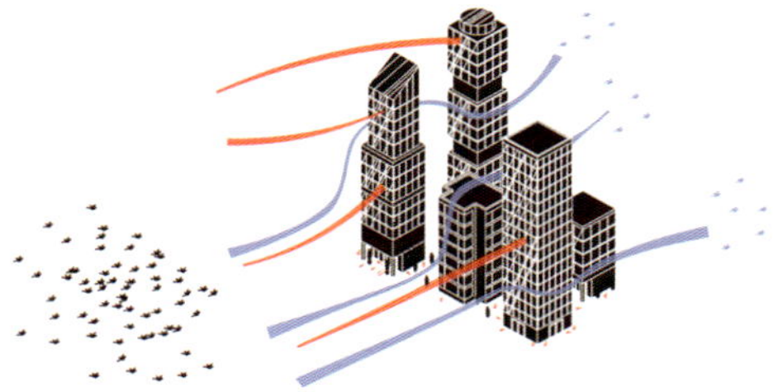

*a*

*c*

*a)Bird strikes on glass façades during migratory season, b) multi-species façade screen concept to reduce bird-strikes and provide bird resting spaces, and c,opposite) fabricated multi-part prototype of façade screen ceramic assembly.*

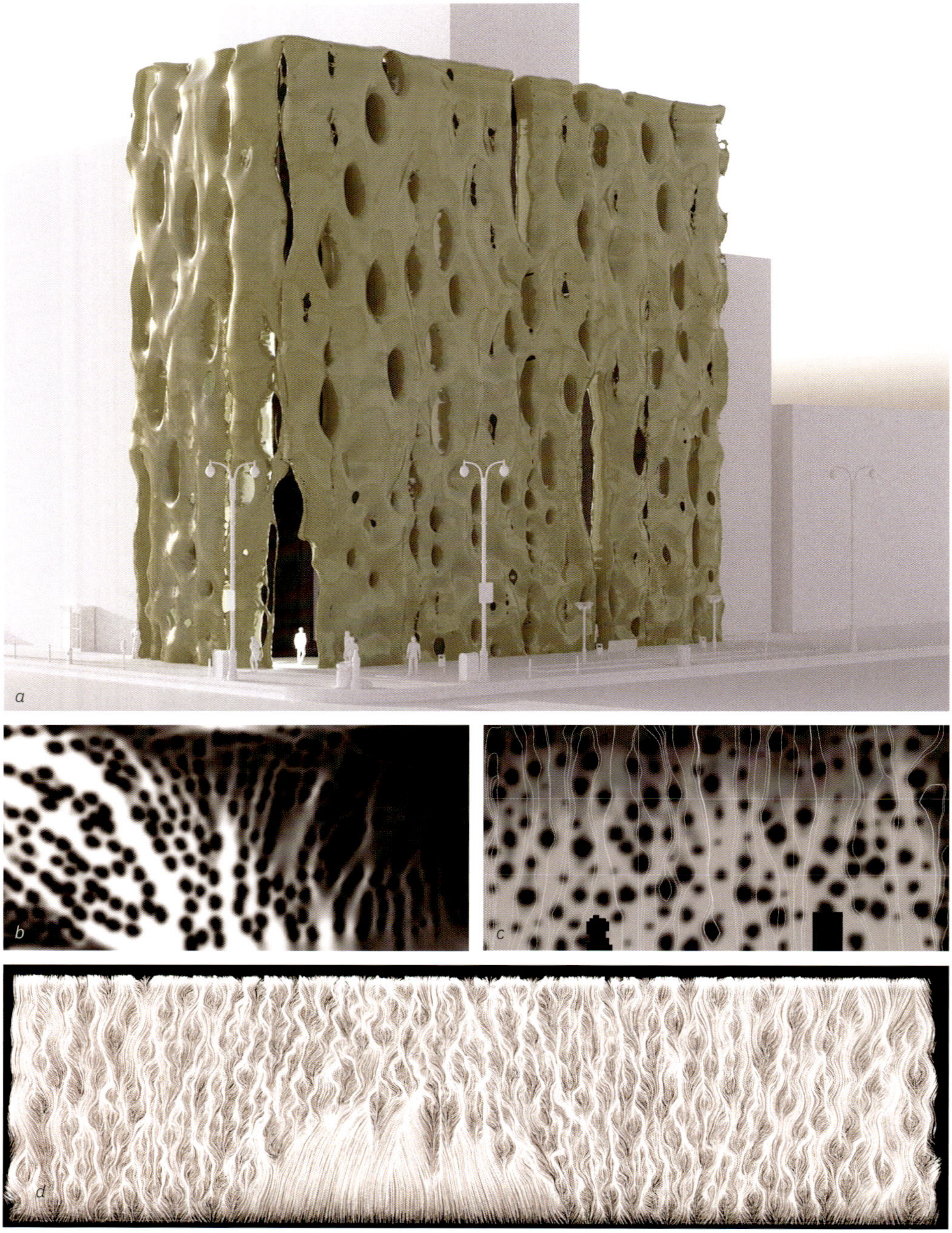


*a)Schematic 3D massing concept for façade screen, generative computer simulations of b)topological structural optimization, c) agent-based vertical path generation, and d) particle-field vertical growth design method.*

*Fabricated multi-part prototype of façade screen ceramic assembly comprised of nine post-tensioned additively manufactured ceramic parts.*

*Hybrid manufacturing process featuring: a)multi-axis, multi-pass additive manufacturing, and b) robotic smooshing (manipulation) of wet recently additively manufactured clay. c) glaze-fired ceramic part, and d) close-up of wet clay directly after smooshing.*

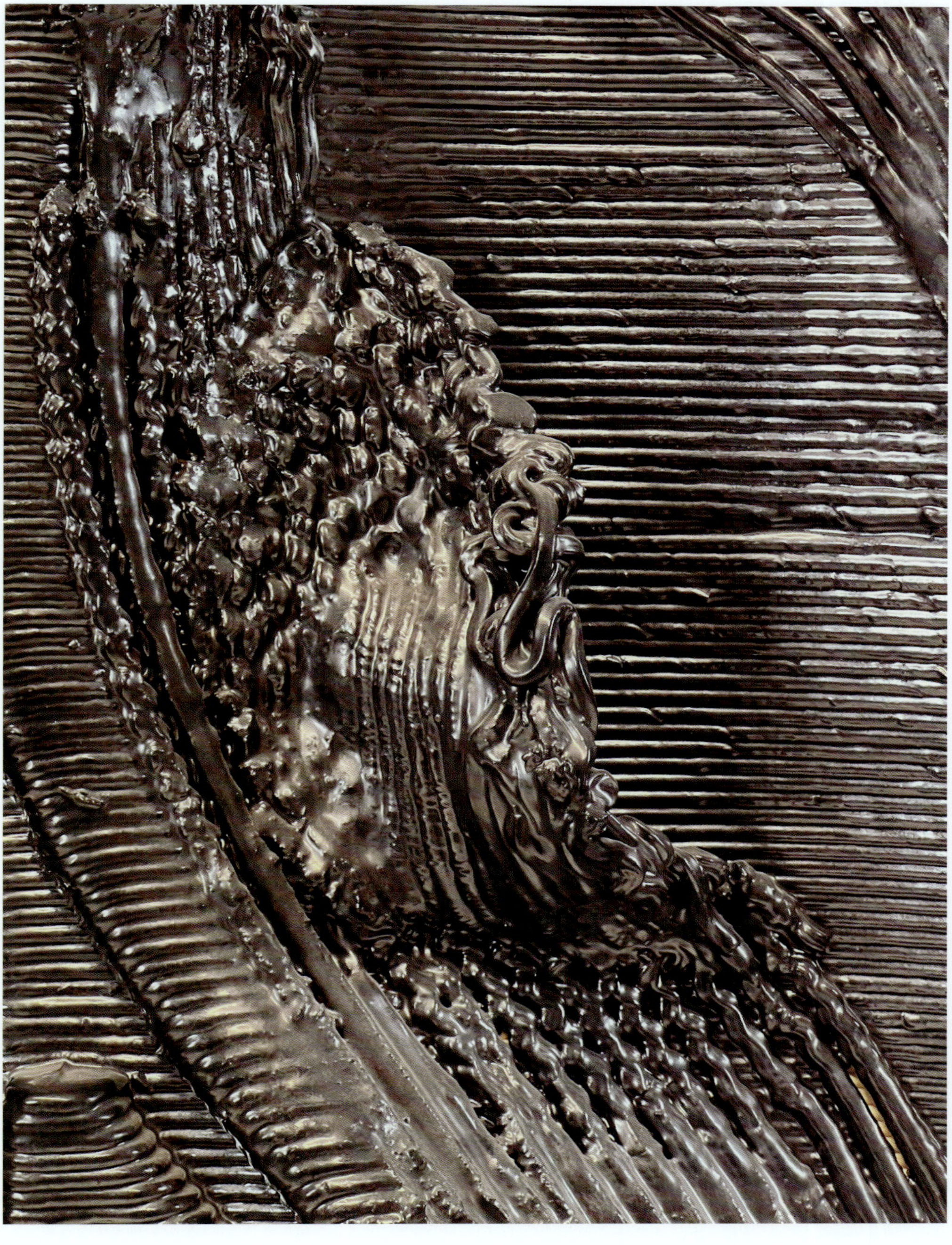

*Close-up of bird resting space that was fabricated using a second vertical additive manufacturing (AM) process ontop of a horizontal AM volume and subsequently smooshed prior to bisque and glaze firings.*

# THERMA-SCREEN

*ADDITIVE MANUFACTURING*

*HRISHI RAJASEKAR*

*INSTRUCTORS:*
*ROBERT STUART-SMITH*
*BILLIE FAIRCLOTH*
*NATHAN KING*
*JEFFREY ANDERSON*

*TAS:*
*HADI EL-KEBBI*
*SOPHIA O'NEILL*
*MATTHEW WHITE*

Amongst other design criteria, façade screens typically negotiate environmental and occupant performance considerations such as daylighting, passive heat gain, alongside user privacy or visual transparency. While ceramic cladding is widely used in multi-story buildings, more recently researchers such as Jason Oliver Vollen have demonstrated their potential to operate as passive thermal capture devices that can be integrated in hydronic heating systems (Vollon 2020). Extending this principle, this project develops a computational design approach to a ceramic façade screen's geometry that seeks to enhance the possibilities of passive hydronic heating within the screen's thermal mass. Following a series of studies that ascertained the mass and thermal conductivity of the façade, solar radiation analysis was performed on a series of prototypical façade screen geometries to evaluate their ability to both increase solar exposure whilst enabling an almost 50% visual transparency through the façade screen. Following this, texture was investigated and demonstrated to enhance thermal absorption by providing greater surface area facing the solar path throughout the day. A fractal-like algorithm was developed that produced increased surface area and variable surface orientation through subdivision and surface displacement. The result leverages the non-standard geometric potential of additively manufactured ceramics to provide a highly textured façade that offers both enhanced thermal performance and unique aesthetic value.

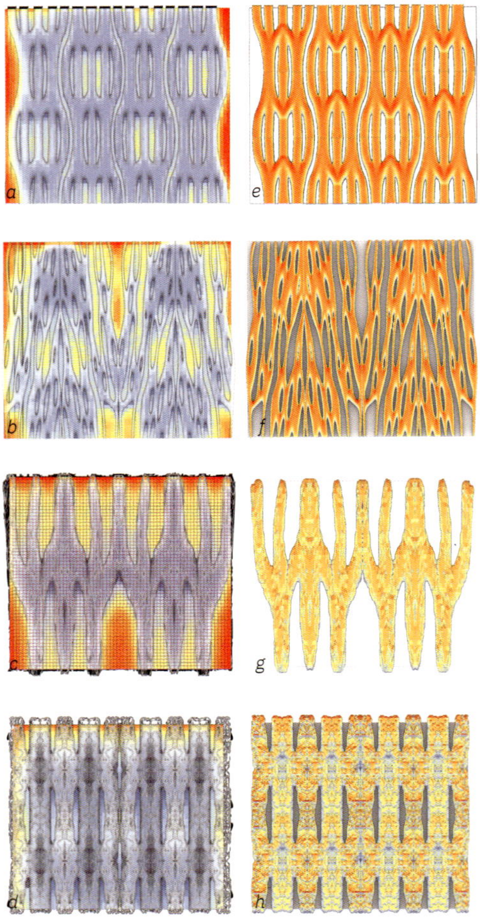


*a-d) Solar radiation analysis of glazing behind ceramic façade screen and e-h) corresponding solar radiation analysis of ceramic façade screen. The façade screen aims to collect heat through passive hydronic heating whilst partially shading the glazing behind. Textured options (g-h) appear to absorb less solar radiation in this orientation. Their increased surface area and variable surface orientation absorb more heat on average throughout the day compared to (e-f).*

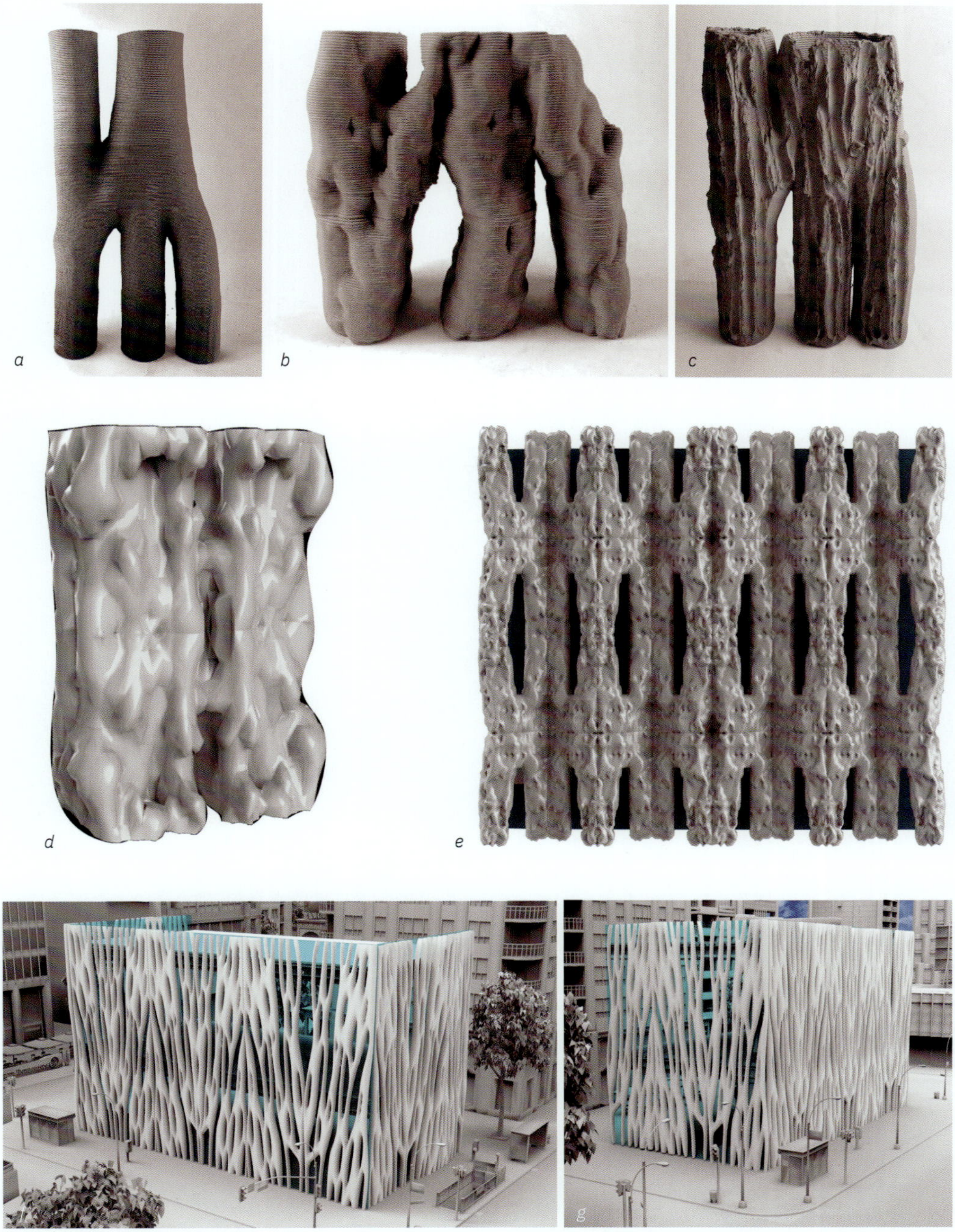

*Additively manufactured clay prototypes: a) smooth, b) volumetrically textured, and c) textured by carving. d-e) custom algorithmic method for surface undulation and fractal texture, and f-g) exterior perspectives of façade screen.*

Vollen, J. O. "Countercurrent Heat Exchange Building Envelope Using Ceramic Components." A+BE | Architecture and the Built Environment (italicized), 10, no. 05 (2020): 1-296. https://doi.org/10.7480/abe.2020.05.5190

# FILIGREE FAÇADE

***DIE-CAST EXTRUSION***

*Adam Blood*
*Nicholas Houser*
*Jeffrey Liao*
*Matthew Ward*

*INSTRUCTORS:*
*Robert Stuart-Smith*
*Billie Faircloth*
*Nathan King*
*Jeffrey Anderson*

*TAS:*
*Hadi El-Kebbi*
*Sophia O'Neill*
*Matthew White*

The Filigree Façade explores an integrated approach to the design and manufacturing of a ceramic architectural façade screen. The research hybridizes industrial die-cast extrusion manufacturing with industrial robot additive manufacturing methods to produce geometric variation in parts by leveraging dynamic kinematic possibilities in the robot production process. The research focuses particularly on the development of custom end-of-arm tooling capable of producing variable extrusions and an agent-based ceramic façade computational design method tailored to the production method's constraints. The current state-of-the-art for robotically extruded architectural ceramics is primarily concerned with an iterative horizontal layering method used to produce a single architectural part. Although this additive manufacturing technique is often referred to synonymously as "extrusion," it bears little resemblance to the industrial extrusion of architectural ceramics, which typically utilizes fixed, geometric die profiles extruding onto a linear material conveyor to rapidly produce whole architectural parts. The research focuses particularly on the development of custom robot end-of-arm tooling designed to accept interchangeable dies and accompanied by a sensor system that provides in-sights into the variables affecting extrusion. By mounting the extruder to an articulated robot arm, a vast number of extrusion tool paths were able to be achieved, enabling significant geometric variation between parts.

A prototypical design for a ceramic architectural façade screen registers this variation through the placement of parts in an orientation where each extrusion trajectory lies in the plane of the building envelope. An agent-based design methodology further highlights the variable porosity and directionality in the façade screen by enabling regions of inter-part alignment and divergence. A multi-part assembled prototype was designed and manufactured using the research methods and demonstrates the possibility of robotic extrusion being applied to architectural façade screens, where the attained variability could support design outcomes being more responsive to context while creating a complex aesthetic language.

*Custom robot EOAT featuring an extrusion die and air-compressor actuated plunger and real-time sensor feedback that helps with precision timing.*

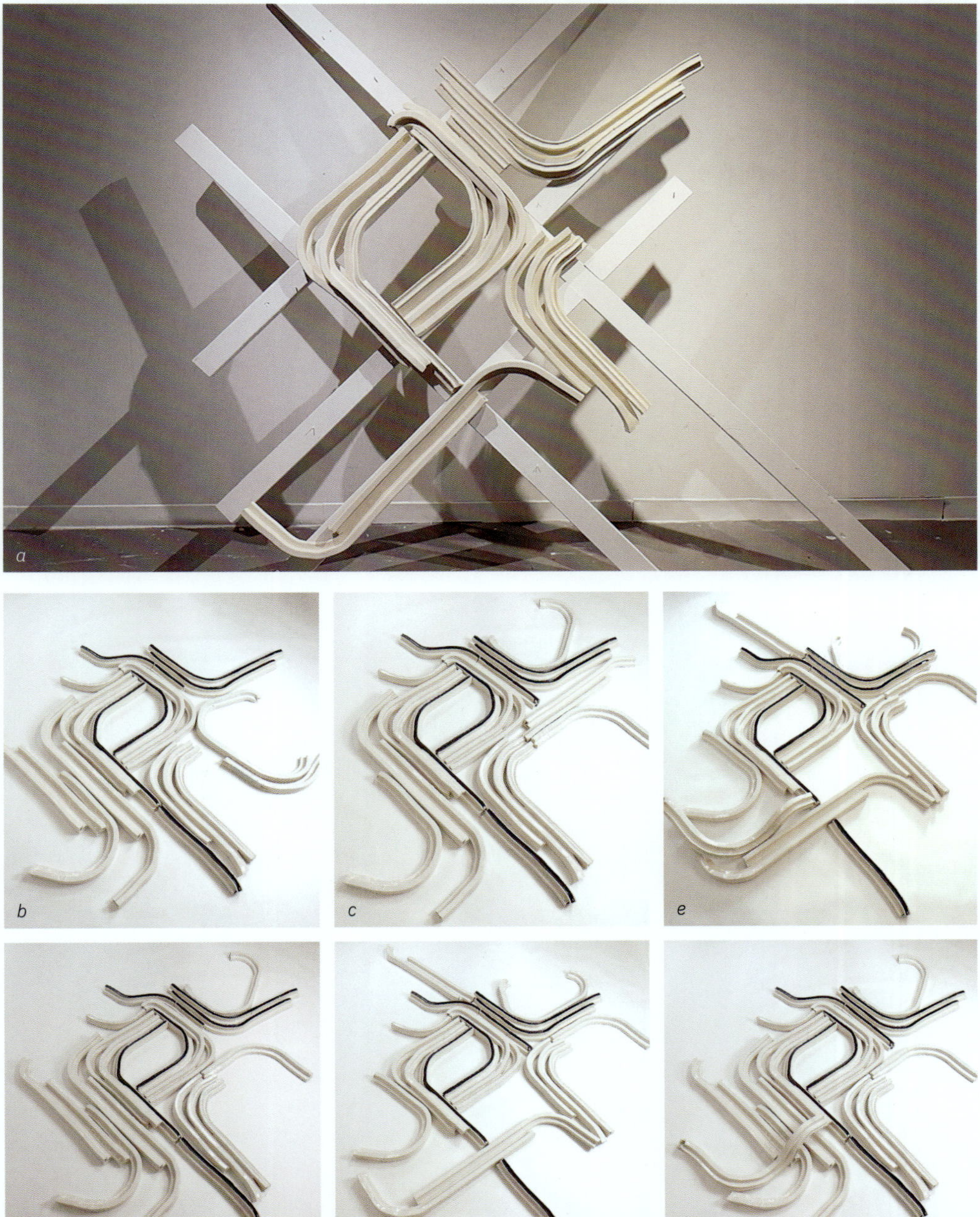


*a) Final full-scale prototype of a façade screen ceramic assemblage, and b-g) extensible combinations of the fabricated parts into different forms of bundled façade screen. Shown post glaze-firing, some parts have different glazed finishes that provide moments of visual reinforcement in the bundled assemblage..*

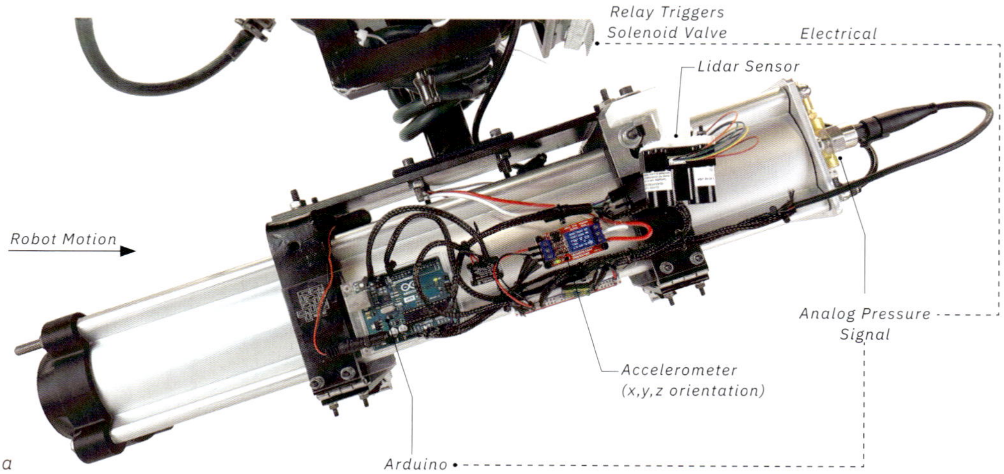


a

b

c

d

e

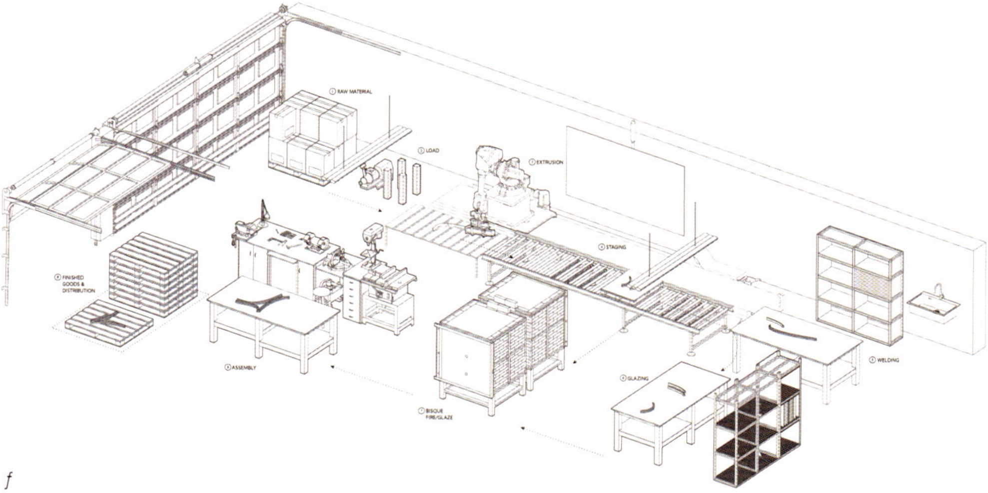


f

*a) Close-up of EOAT featuring an extrusion die and air-compressor actuated plunger and real-time sensor feedback that helps with precision timing. b-e) Design iterations of the extrusion die and f) envisaged micro-factory variable manufacturing set-up.*

*a) Robotic fabrication of one part, b) close-up of die-cast extrusion process, and c,d) multi-agent computer simulations of bundled façade field-condition.*

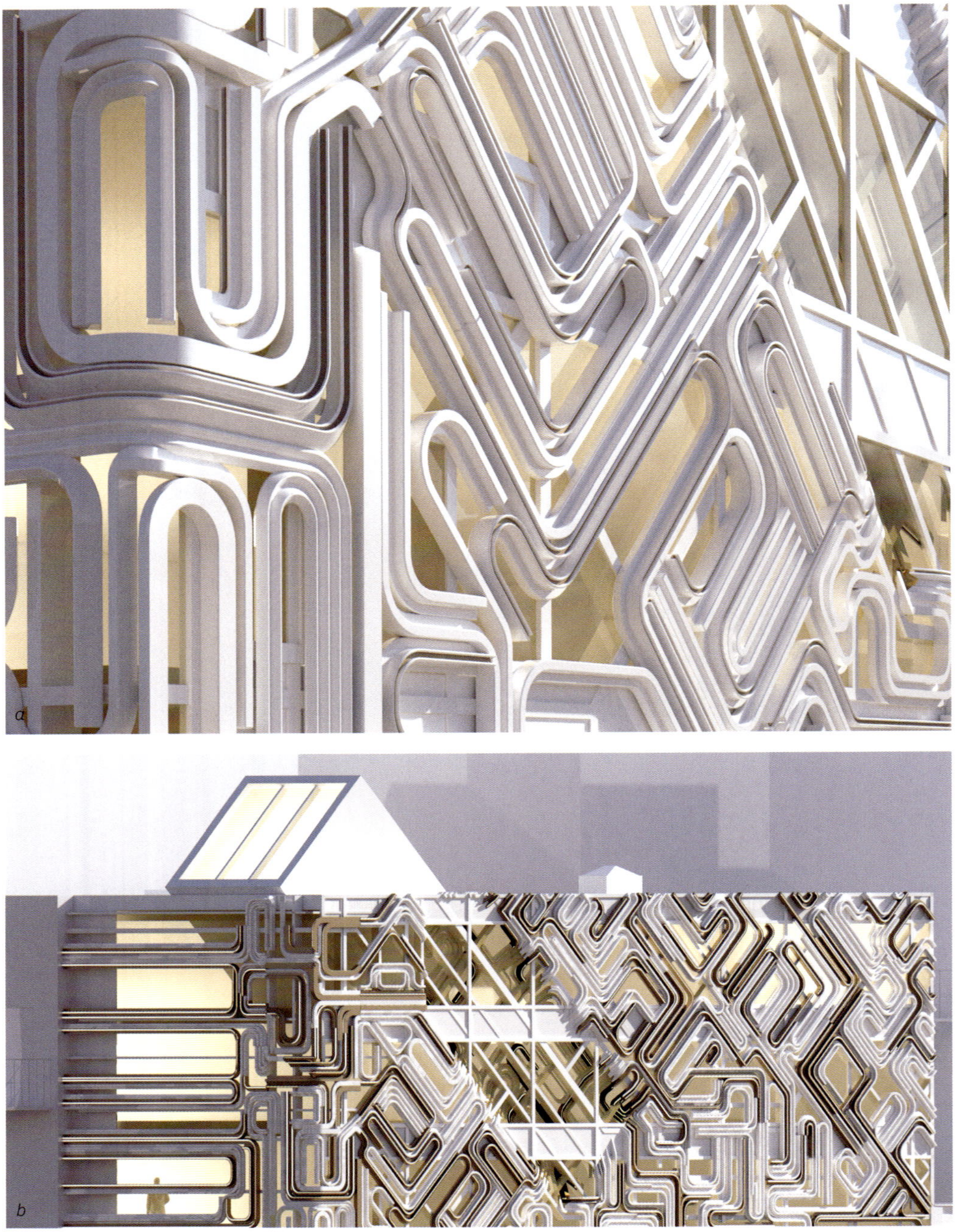

*a) Close-up of bundled façade screen and b) corner elevation of bundled façade screen demonstrating the robotic die-cast extrusion method's ability to produce uniquely formed parts in high quantities, offering exciting design possibilities when strategized across a whole building's façade. The façade is heterogenous in organization and multifarious in its combination of varied parts, offering greater specifity than is typicallly supported by industrial methods of manufacture.*

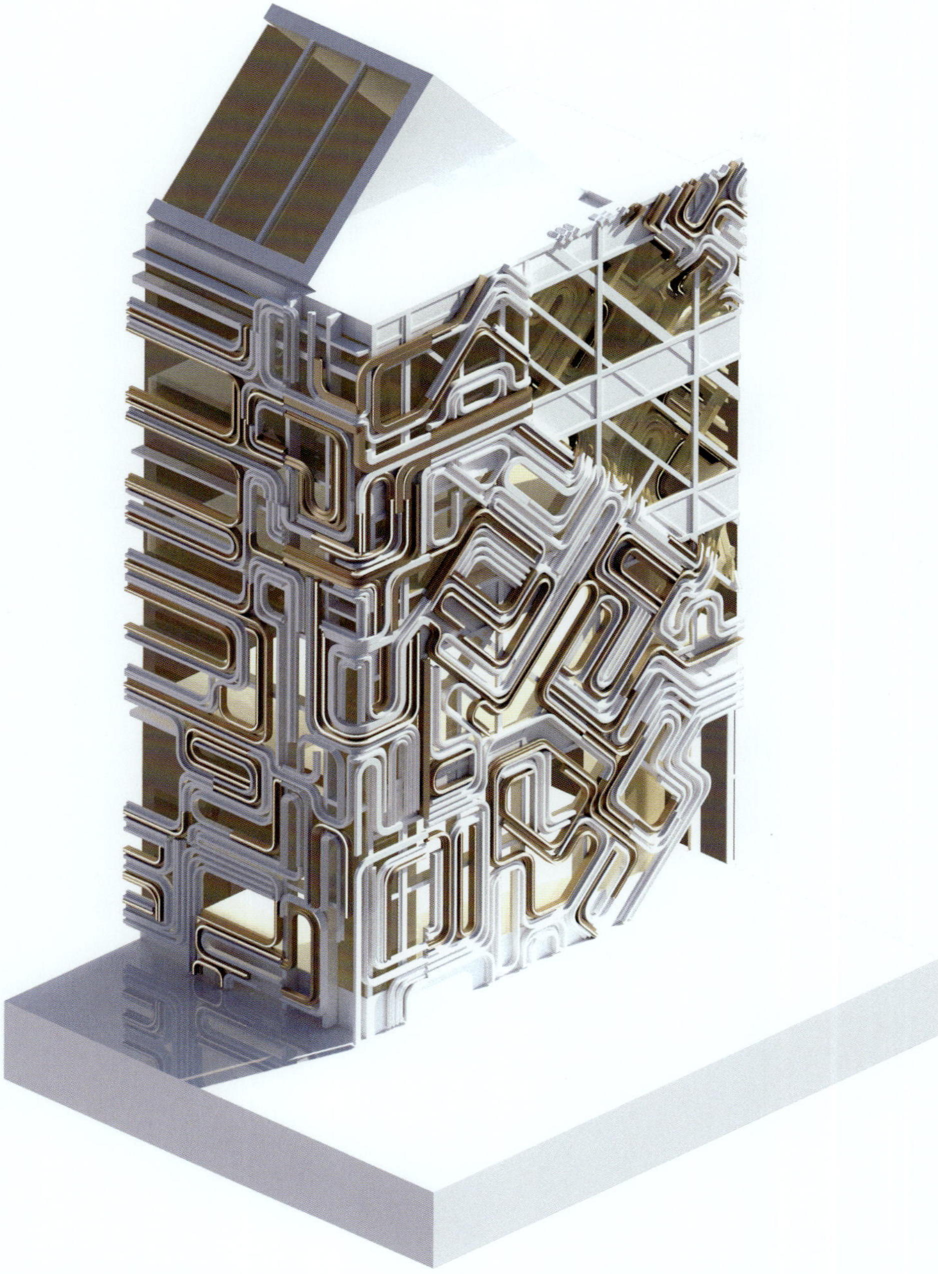

*3D partial chunk of bundled architectural façade screen. Different glazes are utilized to highlight different regions and orientations within the field of die-cast extruded parts.*

# CERAMIC FOREST

### *DIE-CAST EXTRUSION*

*Haohan (Jason) Tang*
*Jingyu (Joey) Luo*
*Shuoxuan (Shane) Su*
*Bentian Wang*
*Pouria Vakhshouri*

*Instructors:*
*Robert Stuart-Smith*
*Billie Faircloth*
*Nathan King*
*Jeffrey Anderson*

*TAs:*
*Hadi El-Kebbi*
*Sophia O'Neill*
*Matthew White*

Extrusion is a well-established industrial production technique commonly used for making architectural ceramic clay parts in high-volume for façade cladding and other applications. In mass production, an auger pushes clay out of a reservoir through a die profile onto a conveyor belt. While the method enables elaborately profiled extrusions, the extrusion and die allow for no degree of variability across the production of several parts. Ceramic Forest explores how robotic fabrication and clay extrusion techniques can be integrated into a variable production process by mounting an extrusion system and die on an industrial robot end-of-arm tool. Experiments exploring fabrication parameters including the clay body water content, die geometry, air pressure, and a robot's motion trajectory informed the development of a hexagonal die-cast extrusion part that supports the ability for several parts to be bundled together to form a spatial façade screen system. The research is demonstrated in the production and bundled assembly of a full-scale partial façade screen prototype comprised of geometrically distinctive parts, and a larger architectural façade screen design proposal. A computational design method was also developed that incorporates the research's established fabrication constraints. Together, these developments illustrate an approach to variable die-formed ceramic extrusion and a computational design approach to the manufacturing method's application in architectural façade screens.

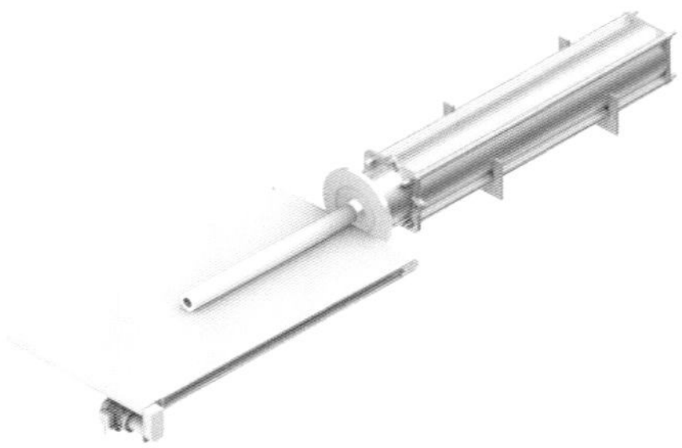

*a*

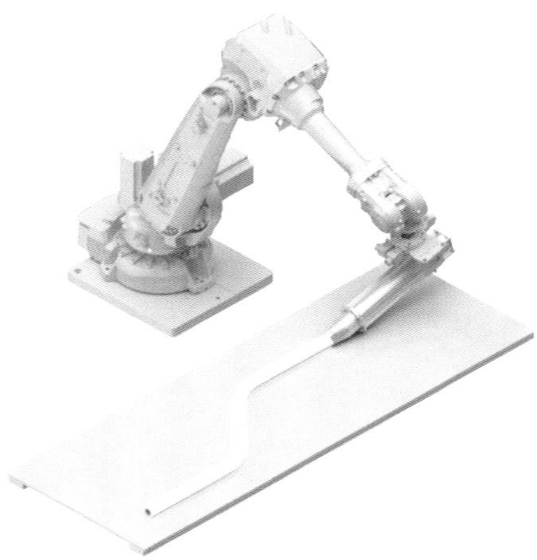

*b*

*a) Industry-utilzed die-cast extrusion does not vary in extrusion path and b) developed variable die-cast extrusion trajectory.*

a

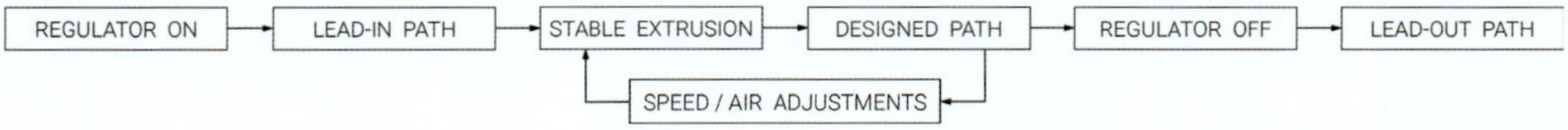


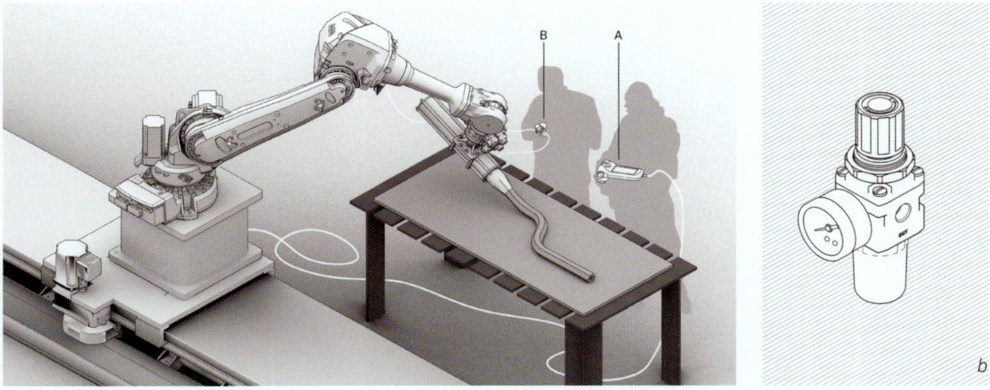


b

*a) Custom-developed die-cast extrusion robot end-of-arm tool (EOAT) and b) robotic work cell setup where each part can be produced with a unique extrusion path.*

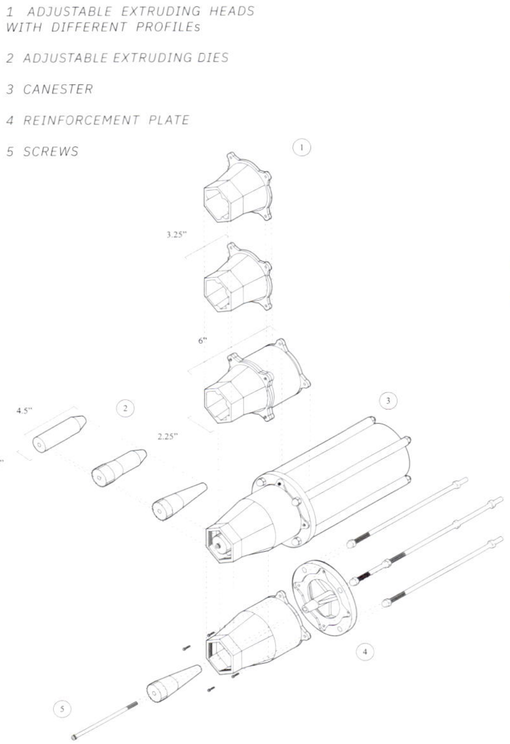


a

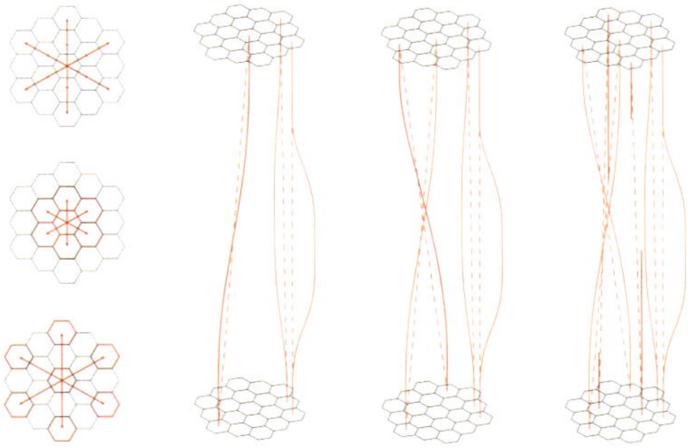

b

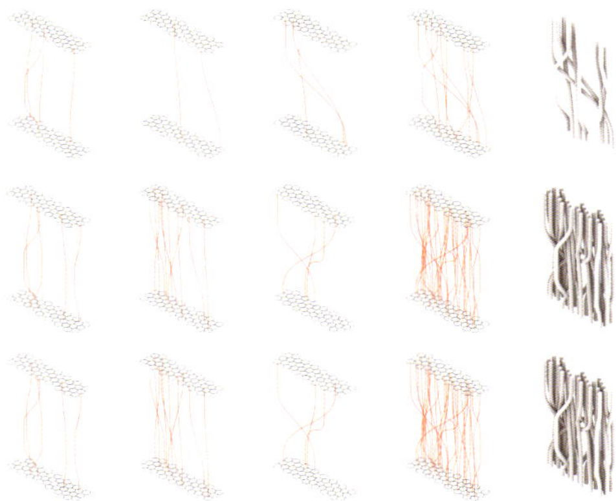

c

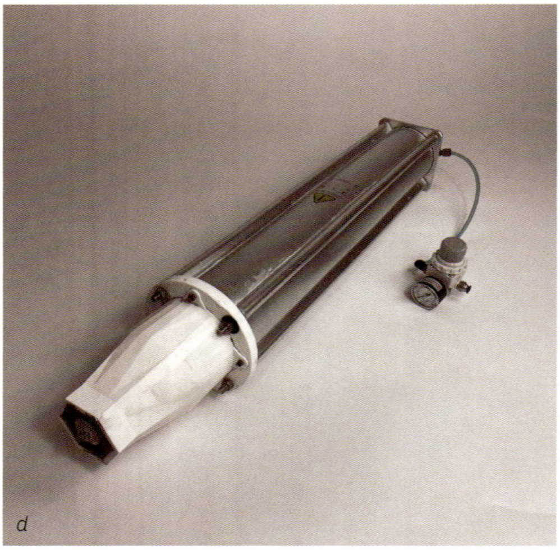

d

e

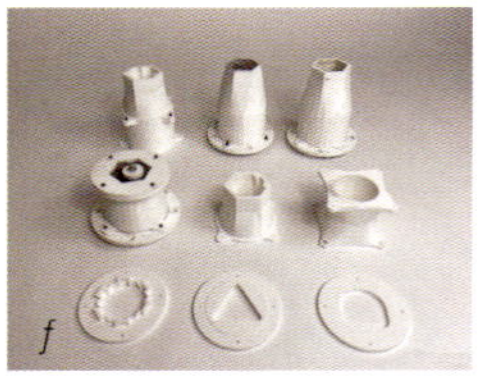

f

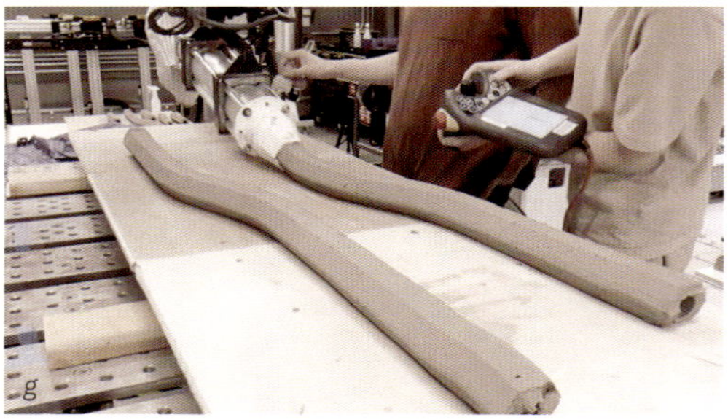

g

*a) Exploded perspective of EOAT design, b,c) developed generative computational design method for producing variable vertical bundling in a vertically orientated façade screen design, d) custom EOAT assembled, e) final die comprises of an internal void section, f) several iterations of the die design, and g) robotic fabrication of two elements with different extrusion paths.*

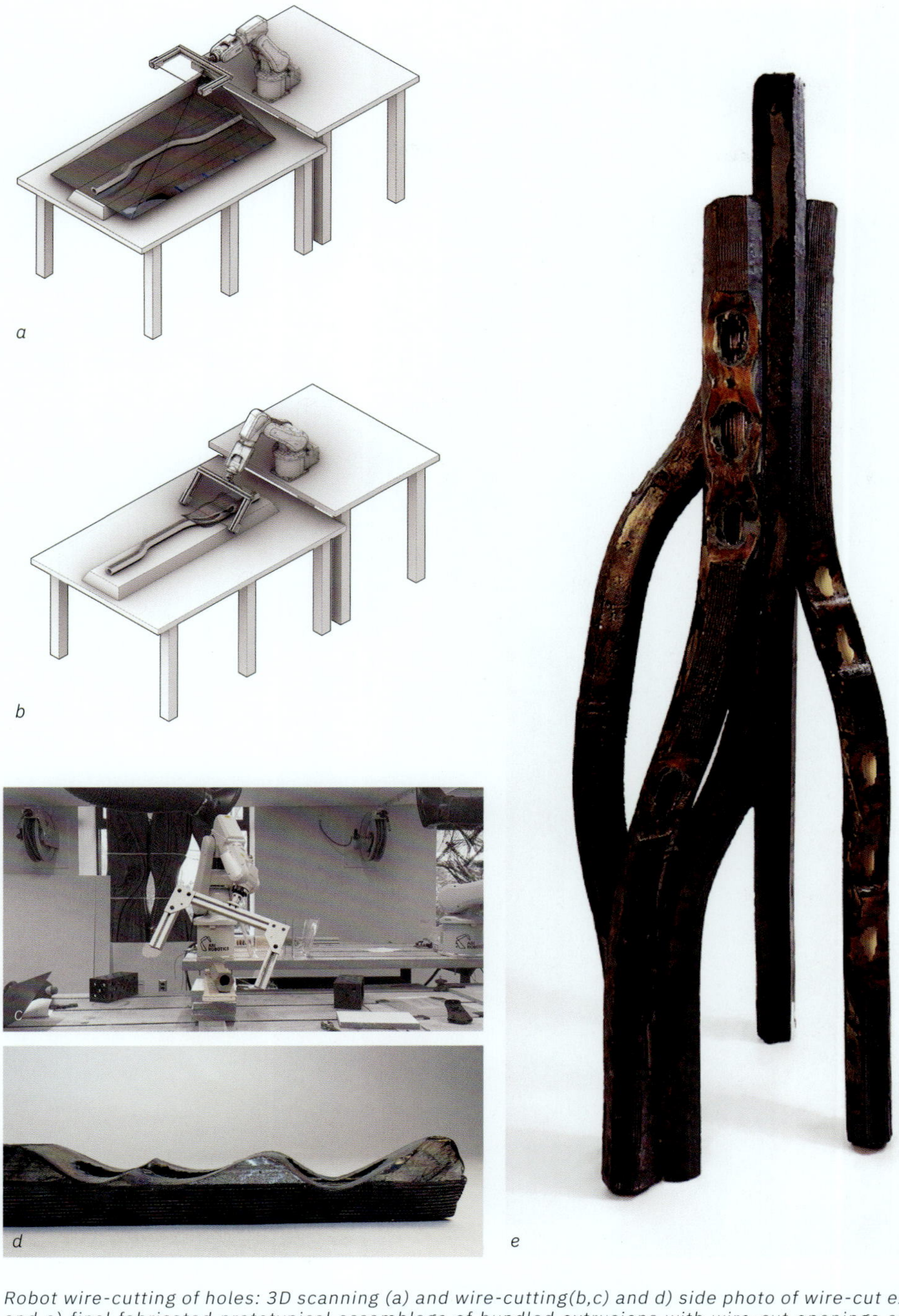

*Robot wire-cutting of holes: 3D scanning (a) and wire-cutting(b,c) and d) side photo of wire-cut extrusion, and e) final fabricated prototypical assemblage of bundled extrusions with wire-cut openings creating a second scale of design effect, revealing the interior of the extrusions.*

*Final fabricated prototypical façade screen assemblage of bundled extrusions. Manufacturing time for all of the extrusions was in the order of minutes—rapid. This suggests the method holds potential for industrial uptake whilst enabling variable deisgn and fabrication possibilities. Post-extrusion parts still needed to be bone-dried, bisque, and glaze fired.*

*a) Exterior perspective of prototypical façade screen design that exhibits heterogeneity in bundle-organization and individual element's geometry and b) interior perspective with wire-cut holes visible in some extrusins. The façade screen provides a degree of privacy and visual dynamism.*

# STEAMED GREENS

***INCREMENTAL FORMING***

*Yuran Liu*
*Riley Studebaker*
*Yuxuan Wang*

*INSTRUCTORS:*
*Robert Stuart-Smith*
*Billie Faircloth*
*Nathan King*
*Jeffrey Anderson*
*Jose-Luis García del Castillo y López*

*TAS:*
*Patrick Danahy*

Steamed Greens speculates on the utilization of urban steam waste to support community gardens and the greening of dense urban centers. A distributed series of green walls are envisaged that attach to existing buildings where they could distribute waste heat and water to support plant growth. As each site varies in geometry, solar exposure, and use, a variable design and production method was required.

A materially efficient 6-axis incremental forming method for architectural ceramic panels was developed that enables custom doubly-curved panels to be fabricated from flat slabs of clay without requiring the production of molds which involve substantial material waste and cost.

Site thermal and geometrical data capture supports a generative design method that enables ceramic green-wall designs to be developed that are adaptive to individual site's solar radiance and desired plant species growth conditions. Steamed Greens advocates for a situated architecture that is custom tailored to site conditions and can recycle existing urban steam waste. Aligning to this, the manufacturing method also supports the production of diversely formed panels with zero material waste from clay offcuts or molds.

*a) Multi-agent branching algorithm is seeded at steam-vent locations and seeks out areas with high solar exposure determined by a solar radiation analysis and b) elevation of green wall.*
*Opposite: Final fabricated prototype comprised of 16 parts that form eight double-sided, hollow panels.*

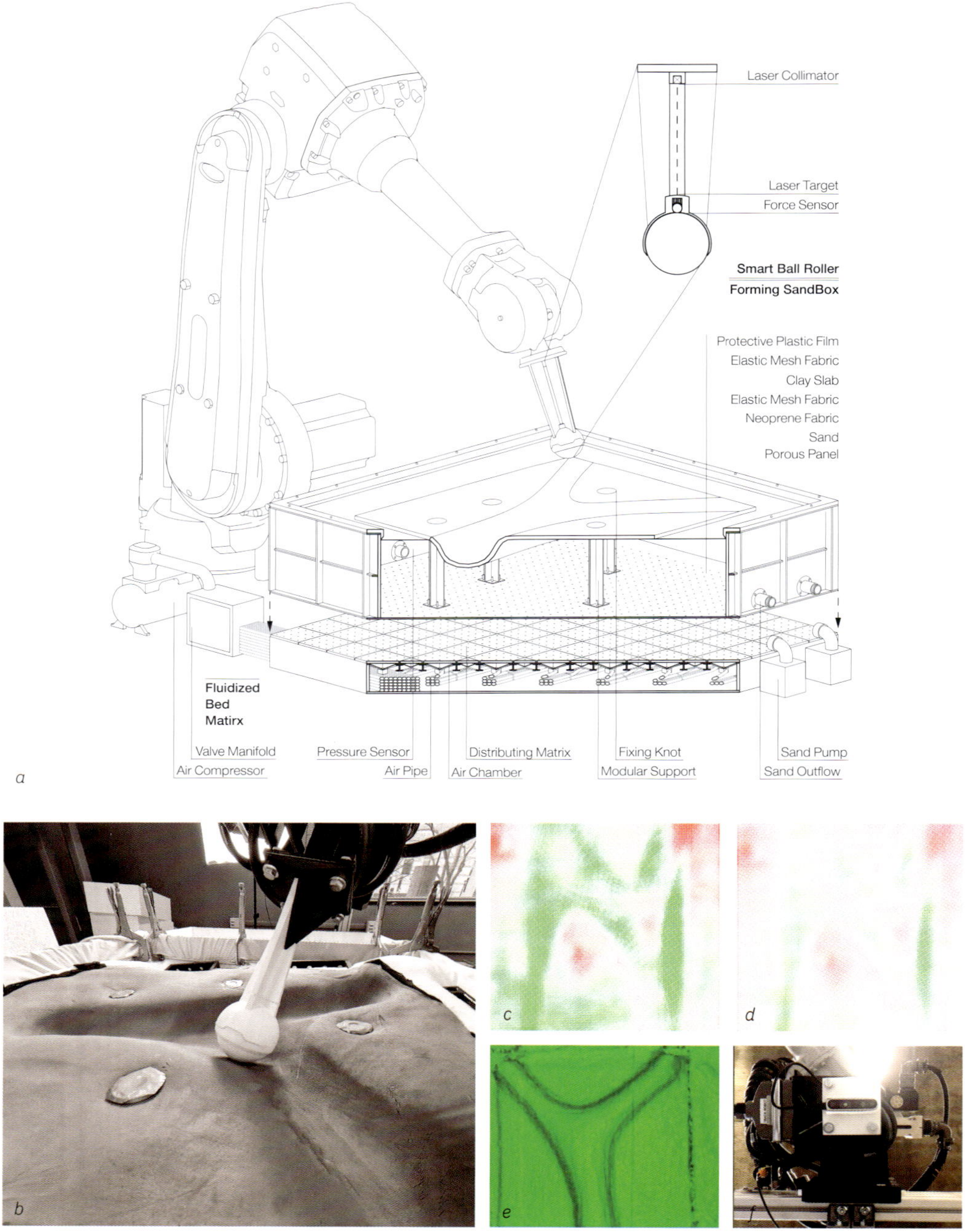


*a) Incremental clay forming method which supsends a sheet of clay in a bed of beads enabling the panel to be supported while it is being variably formed, b) a robot incrementally forms a clay sheet using a custom roller-ball tool that reduces friction and tearing of the soft clay sheet, c,d) iterative forming improves panel geometric results, adapting from an earlier scan (c) to scan (d) while seeking to reduce geometric deviation from the intended panel shape, e) 3D scan, and f) Intel Realsense RGBD used to create 3D scans.*

*a) Two incrementally formed panels are b) connected back-to-back to form a hollow panel; c,e) glazing tests, close-up reveals a material effect of dimpling created from the bed of beads below the clay when the panel is formed. d) perspective view of the prototypical community garden greenwall.*

# LIGHT-ING

### *ADAPTIVE SHEET FORMING*

*Jingjing Yan*
*Siyu Dong*
*Xiangguo Cui*
*Shunyi Yang*

*instructors:*
*Robert Stuart-Smith*
*Billie Faircloth*
*Nathan King*
*Jeffrey Anderson*

*TAs:*
*David Forero*

Building envelope design incorporates a range of light-related global analyses that often inform environmental performance, geometry, or material composition. Extending envelope light-related analyses to include lighting transmittance effects on a building's interior presents opportunities to explore the translucency, porosity, and form of façades in novel ways, yet requires design approaches that are tailored to the production and evaluation of such visual effects. Although ceramics are widely used in façade construction, they offer a wide range of transmittance and deformation potentials that are relatively unexplored architecturally.

This research investigates porcelain ceramics physio-material properties and its transmittance post-firing, together with its relationship to form, material thickness, and porosity, and how these correspond to different interior and exterior lighting effects. Glazed architectural ceramic components are fabricated using a custom moldless robotic manufacturing method that forms variable panel topographies through iterative, localized pressing that both forms and thins clay sheets. In addition to thinning enhancing light transmittance, a real-time adaptive robotic perforation method allows components to incorporate a distributed field of variably orientated holes. Perforations are undertaken to achieve a specified level of light transmittance in relation to computer vision and sensor data in addition to 3D design criteria.

Panels are thus partially designed within a computational approach to 3D building performance and partially during fabrication where a desired light transmittance level arises from the manipulation and perforation of each panel in relation to live sensor data feedback.

The research approach exploits the malleability and translucency of porcelain ceramics within a semi-autonomous manufacturing process that casts manufacturing as a creative enterprise where feedback enables the approach to operate as a variable, objective-driven form of real-time design agency. The design and manufacturing approach was realised in a partial façade screen assemblage prototype and digital façade screen design that embodies variable geometry, light transmittance, and aesthetic effects.

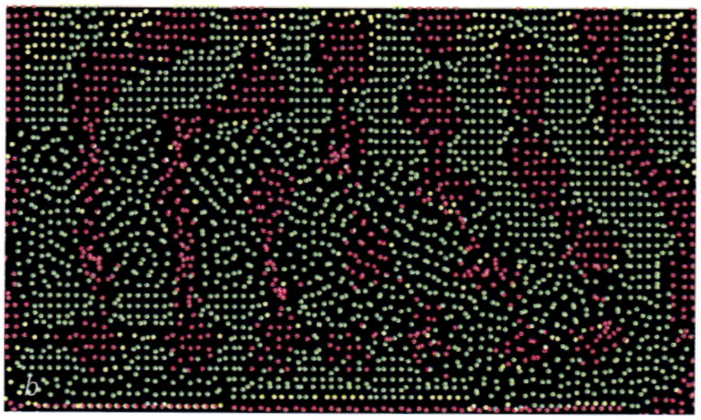


*a) Computational design process: a multi-agent algorithm was developed to determine a spatial distribution of panel pressings.*

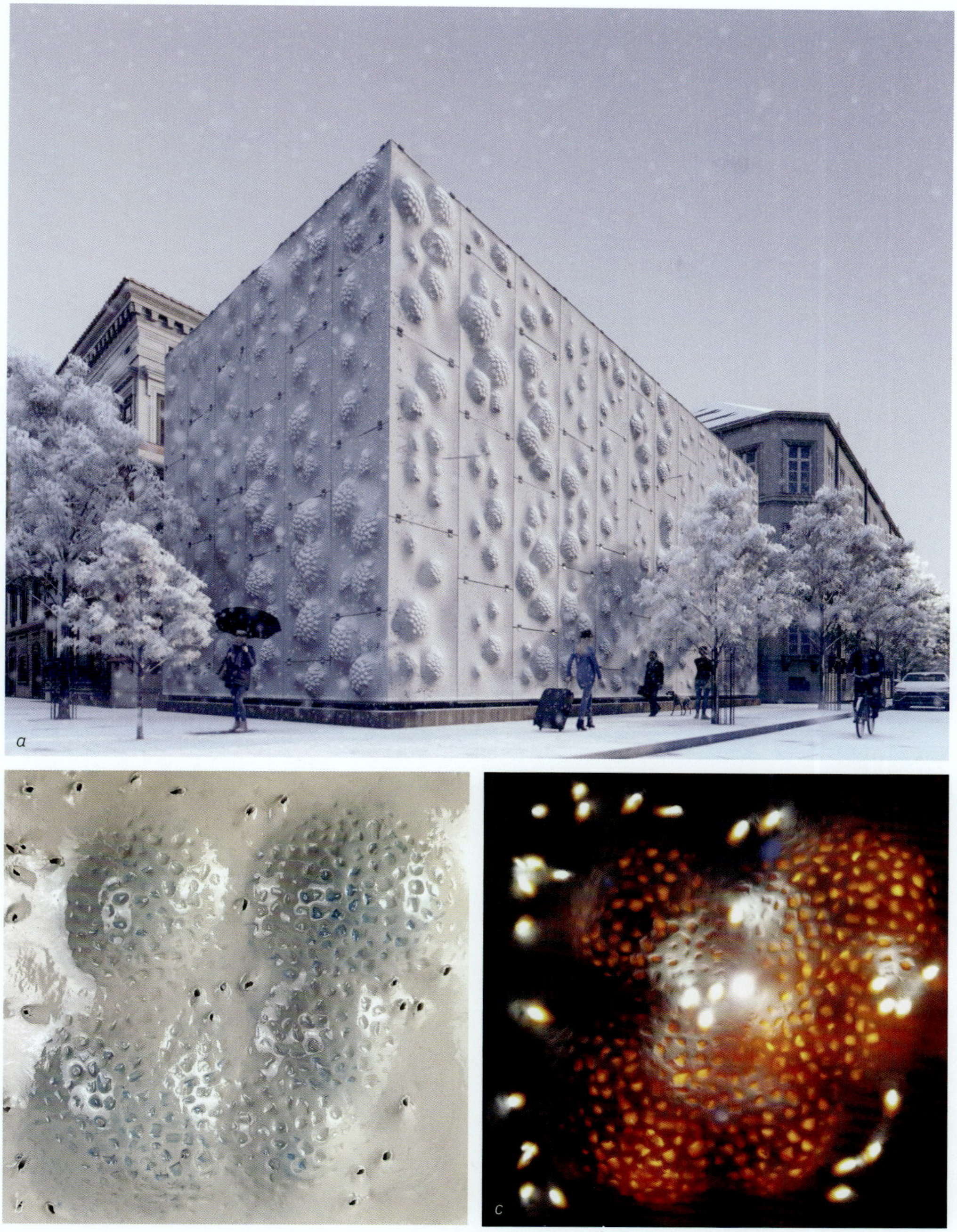

*a) Exterior perspective of prototypical façade screen design that consists of a standardized size of ceramic panels that each contain a unique formal and material condition comprising of variations in depth, form, translucency, and perforation. b) Viewed from the exterior in daylight, light transmittence is not visible, and c) lit from inside (behind) at night varying degrees of translucency and perforation are apparent.*

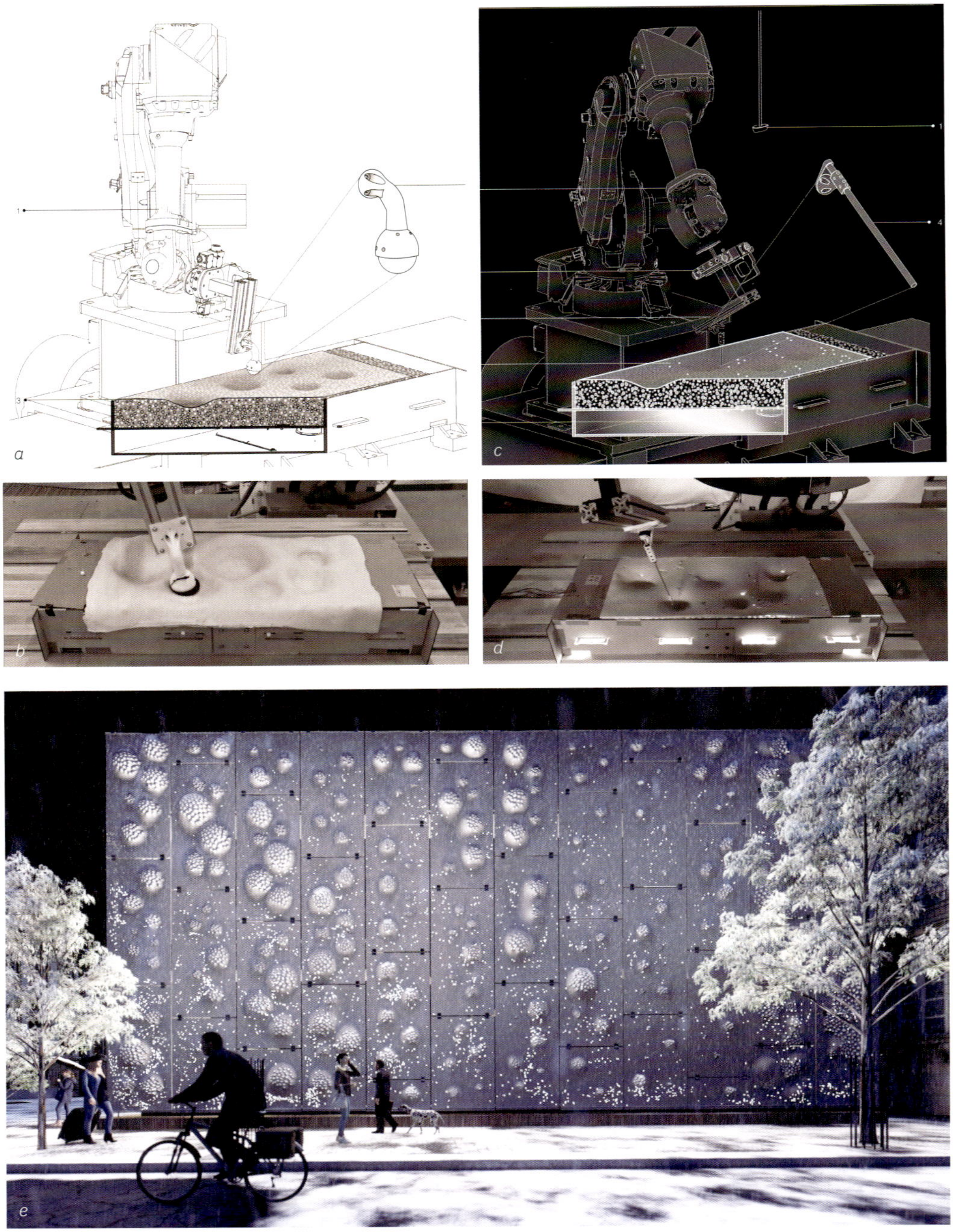

*a,b) A robot iteratively presses on a sheet of clay that is supported by a bed of beads that enables the panel to be suspended but deformed. Pressing thins out the panel, increasing translucency. c,d) The panel is lit from underneath and iteratively perforated by a robot adapting to computer vision data—until it reaches a desired, observable level of light transmittance. e) Exterior perspective of prototypical façade at night, illustrating relationship between formal variation and light transmittence from lit interior.*

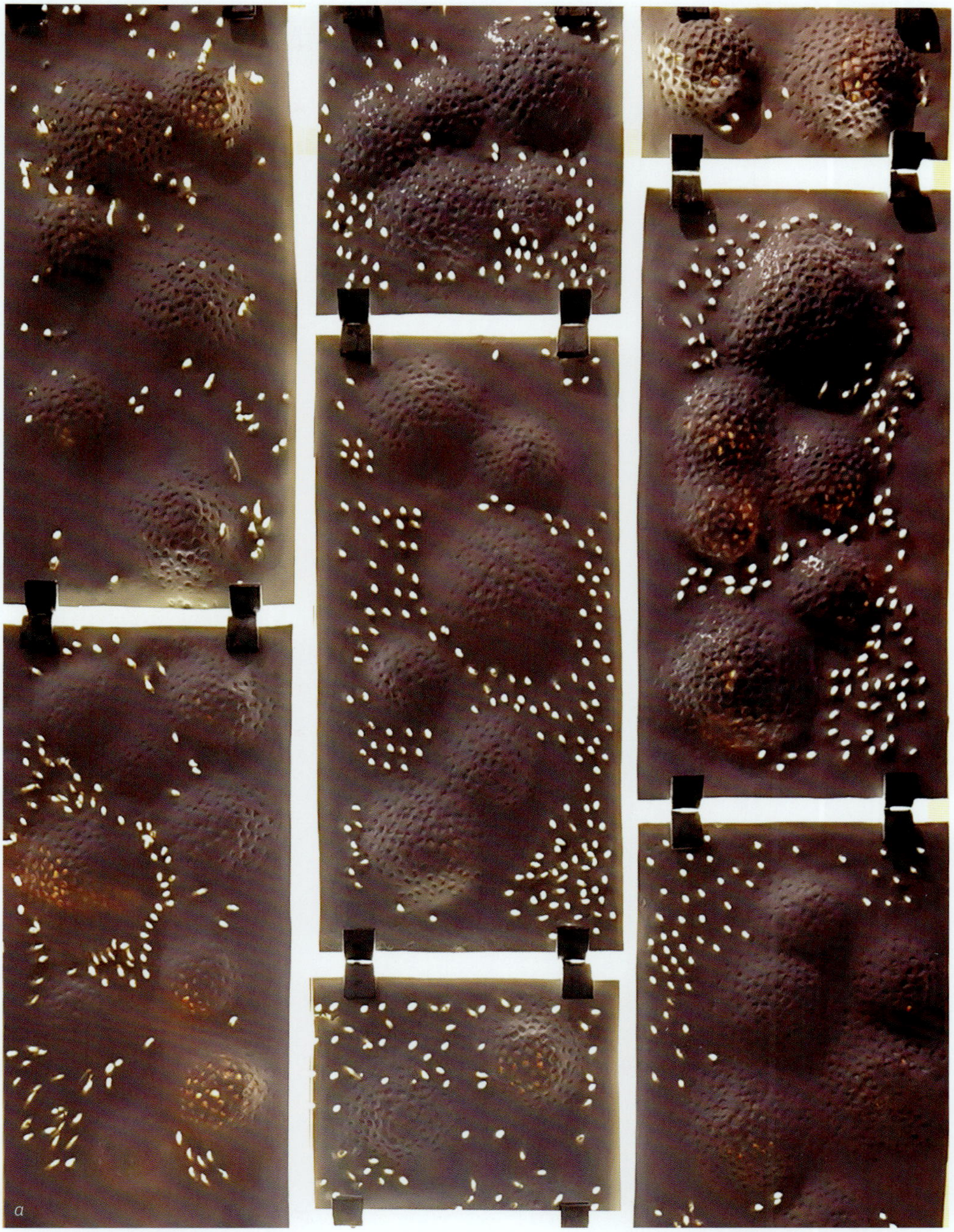

a) Multi-part assembled prototype demonstrating light transluncency and perforation. Each panel was fabricated using a custom robotic pressing and computer-vision adaptive perforation method.

# CAREER TRAJECTORIES

*MSD-RAS Graduates Now*

From the outset, the STEM-certified MSD-RAS program was conceived to enable ambitious career opportunities for graduates.[1] Yet, how effective is the program in supporting such possibilities? To help answer this, MSD-RAS alumni were surveyed from the first three years of the program. Whilst it is impossible to accurately predict future industry and economic trends, this survey data provides first-hand knowledge into the employment opportunities RAS graduates obtained and how they have fared in diverse career trajectories to date.

# WHERE ARE THEY NOW?

## *Surveying MSD-RAS Alumni*

Over its first three years, 46 students graduated from the MSD-RAS program, an average of 15 students per year. These alumni were asked to fill in a brief, anonymous survey. While it is challenging to collect data from alumni who are all busy in their own careers, 27 alumns responded to the survey, providing data that is thus representative of a clear majority of MSD-RAS graduates from this period.

*Results illustrate the exciting potential of the program to empower students to effectively operate in a broad number of employment sectors, in high-level roles with competitive salaries. Specialist skills developed during their MSD-RAS studies also featured prominently in each alumni's employment activities, demonstrating that the cutting-edge MSD-RAS curricula also has real-world value in industry and academia.*

### Diverse Employment Opportunities

MSD-RAS graduates have transitioned to employment in several different areas ranging from academia (teaching, research, or PhD studies), to architectural offices, design-build companies (such as volumetric modular, prefabricated, and platform-based architecture), fabrication companies, real estate development, construction, or tech startups. These diverse roles seemed to be attained quickly with 70% of graduates finding employment within the first month after graduation and almost all graduates finding employment within three months of completing their studies.[2] Also, as MSD-RAS is a STEM-certified degree, most international graduates were able to obtain US-based employment after graduation if they so wished.[3]

### Wide-Ranging Salaries

In the US, first-year graduate architecture salaries in 2023 averaged close to $50,000 per year (Archinect n.d.; ZipRecruiter n.d.). After one year of study in the MSD-RAS program,

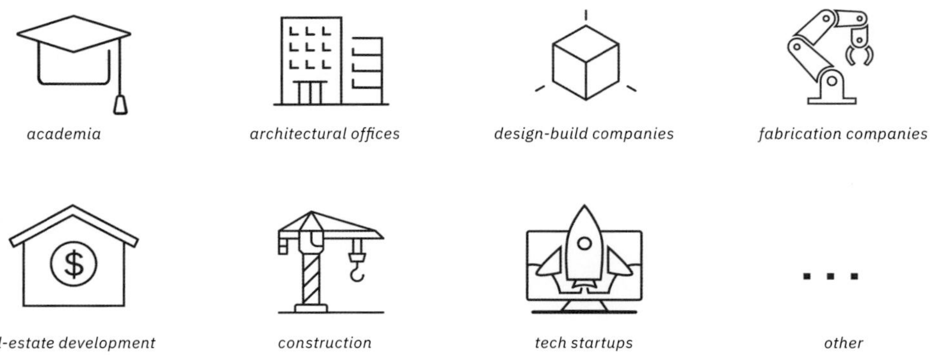


*Different career trajectories of MSD-RAS graduates. These diverse roles directly relate to skills utilization and salaries presented on the next page.*

**70%**

*within the first month*

**90%**

*within three months*

*a*

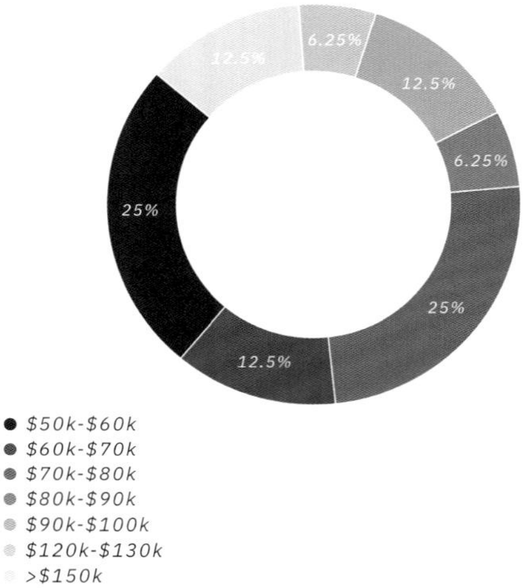


*b*

graduates between 2020-2023 pursuing industry careers received higher salaries that ranged from $50,000 to over $150,000 per annum. While some graduates had prior industry experience, the majority of graduates had little to no industry experience.

## Skills Utilization/Applied in Employment

Students learn a wide range of skills and workflows during their studies. It is pleasing to see the high ratio of graduates who go on to apply these specific skills in their day-to-day jobs. Given the range of industries and roles graduates have grown into, it would be impossible for all graduates to share the same activities in their respective roles. The survey results do, however, demonstrate obtained skills are widely used in graduates' daily careers. While it was not included in the survey, it is quite likely that there is additional indirect involvement with some of these activities occuring where graduates' roles require knowledge in RAS-related subject matter. Several more generic activities were also captured in survey questions but have been ommitted from the figure below (c) such as presentations (51.9%), or 3D rendering/visualization (66.7%), or consulting engagements (22.2%).

## Entrepreneurial Spirit

More than half of surveyed alumni plan to start their own business at some point. Students surveyed completed their studies between six months to 42 months ago. Within this short time-frame, 7.4% already started a business, almost 29.6% wish to start one within five years

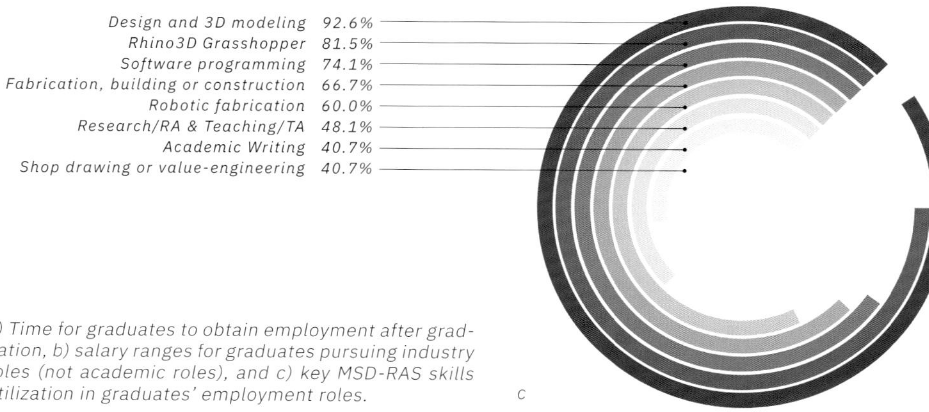


*c*

*a) Time for graduates to obtain employment after graduation, b) salary ranges for graduates pursuing industry roles (not academic roles), and c) key MSD-RAS skills utilization in graduates' employment roles.*

and 18.5% within 10 years. These businesses are likely to leverage MSD-RAS taught knowledge and skills and will collectively help shape the future of industry.

## Design Prominence

It is exciting to see MSD-RAS's graduates excelling in diverse careers and employing the knowledge and skills they obtained in the program in their new roles. Within this diversity of roles and industries, it is also encouraging to see that most graduates are actively engaged in design activities, demonstrating that the program's integrated approach to design and technical work seems to also be evident in graduate's employment roles. The survey does not provide enough information to ascertain whether these roles integrated design and technology by nature of the employment opportunity or due to the initiatives of graduates. Regardless, it is exciting to see design take center stage in such diverse industries and activities. Curiously, perhaps by coincidence—but concievably correlated—the percentage of graduates not engaged in design and modeling is identical to the number running a business.

This survey offers a glimpse into the success of MSD-RAS graduates and the program itself over its first three years. Armed with critical and creative approaches to design and production, MSD-RAS graduates were well-equipped for a diverse range of specialized career trajectories. Looking ahead, we anticipate these graduates will continue evolving into roles that enable them to help lead and positively shape the future of the built environment.

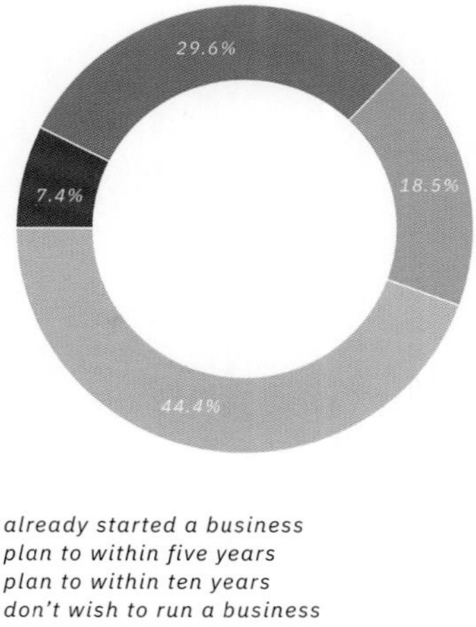


*a*

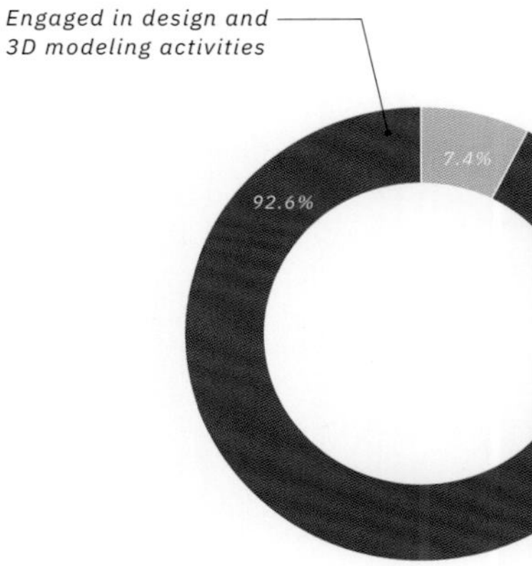


*b*

## Notes

*1. STEM is an abbreviation for science, technology, engineering, and mathematics.*

*2. In the survey, graduates could specify if they found work within 1 month, 2 months, 3 months, 3-6 months, and beyond in similar 3-month blocks. The remaining 10% selected 3-6 months, however, the exact timing for this group of graduates is unknown.*

*a) Percentage of graduates who wish to own a business at some point and b) percentage of graduates engaged in design and 3D modeling activities.*

# CHECKING IN:

*Talking with MSD-RAS Graduates, March 19th, 2024*

**Robert Stuart-Smith (RSS):** *Thank you all for agreeing to meet and share your experiences and thoughts on the MSD-RAS program in this book. This will be a casual discussion where I will provide some questions to serve as prompts, but feel free to diverge from the prompt as you wish. Let's start with why you choose to study in the MSD-RAS program?*

**Claire:** *I was interested in exploring how technology could be leveraged to enhance the human creative process. The MSD-RAS program seemed like a program that would introduce me to that conversation. There were very few programs advertised with the same focus and I wanted a program that emphasized research.*

**Riley:** Agency. I believe that being a trained designer was not enough to bring my designs into the world, and having technical skills in robotic fabrication would provide a platform for my designs to be realized.

**Mahsa**: During my M.Arch graduation, I was fascinated by natural shapes that come from computational designs and wanted to expand my formal language to have the ability to create more aesthetics and explore the future of architecture. While I was exploring, I figured out that all top-notch universities that are working in this realm had a research sector focused on robotic abilities to explore design and fabrication. I was following people who work in this area and found the Kokkugia website featuring work from Roland Snooks and Robert Stuart-Smith, and started to follow them on social media; I immediately found they are both professors in universities and do research at the same time. By exploring more, I discovered the MSD-RAS program supervised by Robert, and thought the program would offer a chance to improve my knowledge and provide opportunities to engage more in research. This is what led me to apply to the MSD-RAS program.

**Dan:** During undergrad, I was working part-time in Penn's Chair of Architecture – Winka Dubbeldam's office. In the fifth year of my BA program was the whole Covid scenario, and it just didn't really seem like getting a job in architecture was going to be really feasible at that time. Someone suggested to me that grad school could be a good option to pursue to obtain new skill sets while the economy recovered. I asked a couple of faculty at RPI what grad programs they thought were interesting. I also asked Winka Dubbeldam, who strongly recommended I go to the MSD-RAS program for its amazing courses and faculty. That was the first I heard of the RAS program. Between that and my excitement for the work shown on the RAS website, I decided to apply to the program.

**RSS:** *What was a typical day in the program like? Did you spend much time in the robot lab?*

**Claire:** We spent every moment we could in the robot lab.

**Riley:** I typically arrived at the lab at 10am to work on robotic coursework, took class in the lab during class hours in the morning or afternoon, had research meetings and extra-curricular meetings in the robot lab during

**Claire Moriarty**
East Lansing, MI, USA
•
MSD-RAS '21
•
Lecturer @RPI

**Riley Studebaker**
Clifton Forge, VA, USA
•
MSD-RAS '21
•
Lecturer @RPI
Director, MetaOrnate

**Dan Rothbart**
New Rochelle, NY, USA
•
MSD-RAS '22
•
Project Engineer, Zahner

**Mahsa Masalegoo**
Tehran, Iran
•
MSD-RAS '23
•
Researcher, Penn

business hours, and continued working on robotic coursework until late evening.

**Dan:** My calendar was a little crazy during the RAS program! I enjoyed this though. Looking back on it, any given day included a class first thing in the morning in the robot lab or other rooms in the Meyerson Hall building, a lunch break (sometimes with a part-time work shift as a robot lab operator), studio (which always met in the robot lab), and designated times to operate the robot arms outside of class time. Although I was perhaps an exception – I was also participating in some student council meetings and events, and emailing/job searching and job interviews! The faculty were helpful in supporting this towards the end of the program especially.

**Mahsa:** Although the program was intense regarding design workload and assignments, most of it was about prototyping and working with the robots in the lab, which made me happy as a maker who has a passion for creating tangible objects.

**RSS:** *How do you compare the experience of the year as a whole to your previous academic studies?*

**Claire:** It was exciting and a whole different approach to creating than I experienced in undergrad.

**Riley:** Very difficult to compare —Covid year, and very different pedagogy than a B.Arch—the MSD-RAS year was very short and intense compared to my B.Arch.

**Dan:** I feel I was rather well-prepared for this accelerated program by my undergrad studies at RPI's School of Architecture. Software and model-making experience led well to adopting the tools of the RAS program quickly. The RAS program was much faster though! The faculty is very impressive, and the mid reviews and final reviews had great guest jurors.

**Mahsa:** *As someone who is tech-oriented and wants to become a cross-disciplinary expert, I noticed a significant difference between working in the RAS courses compared to a B.Arch or M.Arch design studio. In RAS, you always considered the feasibility of fabrication and explored material agency in your design thinking, even at conceptual phases of a project. Fabricating a one-by-one prototype as a proof of concept also allowed me to push the boundaries of fabrication techniques in each project.*

**RSS:** *What do you cherish the most about your experiences in the program?*

**Claire:** I will cherish the friends and colleagues that I met in the program and the radically different career this degree led to.

**Riley:** The long nights of working with robots and good friends.

**Dan:** The networking with people at Penn! Making connections with faculty, students, and their peers was very rewarding. There was a lot of extracurricular activities to participate in this regard.

**Mahsa:** While I was studying and immediately after graduation, I had a chance to work as a researcher and designer on a DFD (Design for Disassembly) project that operates primarily through computational and robotic-based tasks, diving into automated fabrication, where I have gained more experience and knowledge in working with 6-axis industrial robots. This has enriched my insight into the limitations and necessities of these kinds of processes.

RSS: *How did you find your colleagues – any strong friendships or peers you learnt a lot from?*

**Claire:** I greatly respected and learned a lot from a lot of my colleagues in the program. Riley and I learned a lot from each other during long hours in the lab both during and after graduating from the program.

**Riley:** Colleagues were fantastic – I have professional and lifelong connections with several students and TAs from the RAS program.

**Dan:** Working together on the final project brought my group close together. Just knowing what we wanted to accomplish and the steps we needed to take to get there, we were working very closely in the second semester. There was a huge push to the final, and the effort to get there is something I'll certainly remember.

**Mahsa:** I have some colleagues in the lab right now who have more experience than I do. They are genuinely helping me learn more about robotic operations and the mechanical aspects of robots. On the other hand, working with Robert Stuart-Smith and Masoud Akbarzadeh, who are both my co-advisors in the current research project, has broadened my insight into the technical and structural side of real-scale projects that engage with robotic fabrication.

RSS: *Do you think the spring semester's integrated approach to the thesis project across all courses was beneficial?*

**Riley:** Yes—even more integration would have been appreciated—a lock-step approach to the education, where possible, made learning so many things so quickly more achievable – being in the first year of the program, this integration was in its infancy, but the program has become much more integrated now.

**Dan:** Yes, I think this approach was valuable: it is synonymous with the way projects are handled where I work at Zahner. Many teams of people touch every project from concept to fabrication to shipping to installation. The right solution for the project considers all of these factors and manages them. It was a large challenge during the RAS project though and frustrating at times ... trying to do a project that was interesting in terms of architectural design and computation approaches in studio while being interesting in terms of scientific research and writing, and having to also include the development of an awesome robot tool, and the production of a large, well-put together final prototype. It was really difficult to do all of that the same time. I think if the program had a little more room to relax on the integration, it may be beneficial, maybe even enable individual students to specialize a little more, but still be prepared for a high level of integration.

**Mahsa:** Integrating all courses in one project is a super fascinating idea, and pedagogically it is quite different from many design schools that try to unsuccessfully do it all in the studio. Still, until now, I don't believe there is any one perfect model. It was though a joyful and helpful experience, especially for people who want to continue their academic journey by pursuing a PhD.

RSS: *What are you doing now? Did studying in RAS help you obtain that opportunity or excel in it?*

**Mahsa:** As I mentioned, I am currently working as a design researcher in the Weitzman School of Design at Penn, jointly employed by the Autonomous Manufacturing Lab (AML-Penn) and Polyhedral Structures Laboratory (PSL), which has given me additional time to engage more deeply with computational design and robotic fabrication. This work is also directly related to my MSD-RAS studies.

**Dan:** I now work as a project engineer at Zahner. My role involves coordinating the design, engineering, manufacture, shipping, and install of custom architectural metal systems. Balancing the needs of the client, the engineer, the architect, the shop, R&D, and more. Custom processes and automation are paramount in order to deliver accurately, affordably, and on time, in a collaborative manner. There are also a lot of digitally fabricated jigs and tools. The MSD-RAS program was certainly helpful in obtaining this job opportunity and enabling me to perform well in the role. It was a nice, smooth transition from the RAS program. Particularly in terms of workflow, like at RAS, we use Python and Grasshopper at Zahner fairly often for robotics and also for general parametric façades and the automation of production tasks. There is a lot of exciting work here!

**RSS:** ***Zahner is a very successful, industry-leading leading company. Do you want to maybe tell us a little more about what Zahner does? And why it's so exciting to be working there, too.***

**Dan:** Yeah, I would probably say that Zahner is the leading architectural metals company, it is also experienced with custom solutions, as well as offering some standard products. Whether custom patinas on sheet products or large scale architectural projects like façades and façade systems. We use a lot of automation in the engineering group where I work to automate shop drawings and 3D models. And and we also work with a lot of digital fabrication for jigs and tools to help automate fabrication and field installation processes and just sort of augment the skills of our union workforce there. The shop and the field are almost entirely comprised of local members of the Sheet Metal Workers Union, which is one of the oldest unions in the country, and it's a really interesting relationship we have with them in the company.

**RSS:** ***Did you say that the Sheet Metal Workers Union is in both your factories and also on site? So they're bridging both those roles?***

**Dan:** Yeah. Its quite helpful on some of these projects, especially when you're out-of-state and you can connect with the local Sheet Metal Workers Union in that state to find some experienced workers to help install projects. This has been a very long standing relationship at Zahner, which has been a family company since 1897. And so, you know, when our research is related to construction automation, it's almost always sort of leveraging what that experience and aiming to augment what a union workforce can do, to kind of bridge the gap between these two different ways of working.

**RSS:** ***Having had an architect background before RAS, how you see yourself in your current role now relative to those different hats you've worn at different times?***

**Dan:** Yeah, I think the tool-making in the RAS program, which involved making custom tooling and jigs, was invaluable to what I do now. In RAS, it was a one-to-one sort of working to develop a custom-focused process, while in some of our custom projects at Zahner, work has to operate at larger scales of production yet frequently needs some digitally fabricated, custom parts. For this, I can work on it directly, and in a dedicated area where it doesn't have to be a shop work order – so its quite empowering. For the architect hat – Zahner develops relationships with architectural offices, and, for instance, we ran a workshop with Perkins&Will establishing a digital workflow so that we could be involved earlier in their design process. For this, they brought about 15 key designers from several of their offices to our shop. This was very different from a Zoom call working on a mostly finished design with them; we were developing small ideas in prototypes, and my conceptual training in architecture has been helpful in thinking abstractly about these types of design problems and in communicating ideas with architects.

**RSS:** ***Dan, how was your experience looking for work like? You started at Zahner in some kind of short term role, didn't you?***

**Dan:** Yeah. I was thinking creatively about looking for work while I was in the RAS program and trying to think with that transferable skill set that I could get many different types of roles, and I was applying for lots of different companies and asking the faculty if they could put me in touch with people, and they were very helpful, although it just so happened that I secured an internship that wasn't directly connected through any RAS faculty. It was a three month starting position that was basically like an extended job interview. A couple of weeks into the internship, I got to see a lot of very interesting work that I hadn't seen before, or hadn't been published yet from Zahner, so I told my boss that I didn't need three months to know that I wanted to work there and asked if he thought it was a fit from their side, I could work full-time. After about a month, I was hired full-time, and I still find it, of course, very, very rewarding.

**RSS**: *Clair and Riley, what are you doing now? Did studying in RAS help you obtain that opportunity or excel in it?*

**Claire:** I'm currently teaching at the Rensselaer Polytechnic Institute (RPI). Studying at Weitzman as well as teaching and conducting research with RAS faculty led to this opportunity. Riley Studebaker and I are currently leading efforts to remake RPI's robotic fabrication space and have had great success engaging the student body through workshops, studios, and research opportunities. We're both full-time. We both teach two classes, a first year design studio and some sort of seminar every semester. So I think we're both up at a campus most days. Studio three days a week in in the afternoon, and then usually, we're working in the robotics lab on Tuesdays and Fridays. We also have our own independent research teams focusing on different things. Riley can talk about his, but mine's working with edge detection and updating toolpaths in real time. And then I did some other work with a different faculty member to help them on an incremental mental formed part for a light project. I'm also doing a Arduino seminar next year on responsive architecture. It's been great. It's RPI's School of Architecture 200th anniversary this year. And one of their initiatives, of the president who started almost two years ago, is to push robotics in the school. So we've been getting a lot of support in that way.

**Riley:** Yes! I am teaching robotic and computational architecture studios in RPI's School of Architecture full time, as well as running a business that 3D prints façades in my spare time. There was a direct and strong path made possible from the RAS program to my current successful career trajectory. We are going to be installing our first project in West Philly. Depending on how you categorize it, it is arguably the first 3D printed ceramic façade in North America, and it just happens to be that the project landed only a mile away from Penn. It is a collaboration with an artist. Although we basically have carte blanche, where the artist essentially said, would you like this canvas? And so there's their artwork that is more of a traditional mural around this wall, and then there will be this 3D printed façade kind of nested into the mural.

**RSS:** *Sounds fantastic. That is likely to be the first built, robotically fabricated RAS Alumni project from a RAS Alumni business. Claire and Riley, you have done a lot of interesting things in very short span of time since graduating from RAS, right? From working at Jenny Sabin Studio, Claire, to Riley working in our ARI Robotics Lab. You also both worked on the Deep Relief project with RAS Faculty Andrew Saunders, co-presenting the project at ACADIA '22 and at the Post-Digital Neo-Baroque conference in Austria. It'd be great to hear a little bit about those experiences and maybe, why you chose to do those things after RAS.*

**Claire:** I was very interested in Jenny Sabin's Office. Her lab's research was often referenced in my RAS project during my studies, especially her clay printing research. I was excited at the potential to use some of my newfound skills in practice to build at larger scales. It was an exciting opportunity in that way, and I think she was excited too – to see what was going on in the new RAS program as well. Working with her, I worked on her Google installation, mostly using the computational design skills we learnt in the program to simulate fabric material behaviour within a design and production workflow to work out what shapes needed to be cut while accounting for stretching. So even without the autonomous aspect of the RAS experience, fabricating at scale and prototyping as opposed to making architectural models was really important in that job and working with an understanding of materials and tolerances at a real scale, and what else?

In the responsive installation for the Ontario Science Center, skills I learned in RAS with Arduino's and microcontrollers helped me to be able to design and build such systems. I also briefly contributed to a 3D printed metal project, super cool.

**RSS:** ***Awesome. Do you want to talk a little bit about your experiences working on the Deep Relief wall with Prof. Andrew Saunders and architects of record Erdy McHenry Architecture? I think it's the first project from RAS faculty that involved RAS alumni to be built outside of the program. It's located in a real library in a local community near Penn. You obviously got to apply a lot of knowledge you acquired in the RAS program there, and the project was directly tied to one of the design courses you did. Did you also learn things through that experience itself? What was it like to go through that and to be able to present the project in a couple of conferences too?!***

**Riley:** I won't sugarcoat it. We had to figure out a lot as we worked on it; it was a very challenging project to both design and manufacture in a short amount of time. I think that from the experience, for me, what stood out was that there were these deadlines and certain deliverables, and at first assessment I often felt that there was no way that we can deliver those things. So how do you strategize and work towards delivering something like that? Outside of robotic fabrication, sometimes the project just needed to be pushed in a very realistic way like any other project. I do think, though, that it was extremely successful.

**Claire:** It was great to experience the realities of realizing something robotically fabricated, instead of just fabricating prototypes. Fabricating a full thing has issues like, how do we put this in a truck or what off-the-shelf hardware can we acquire? And how can we make this work with such wacky forms? How can we understand both the advantages and also the challenges of doing this sort of design and fabrication process? Like, okay, the tolerance isn't 100% perfect. So let's change the design so that the tolerance is less visible. It was honestly a lot less automated than you would think if you didn't know about this sort of fabrication – there were around 240 different programs developed to make it automated. So it was that sort of experience where it's not as apparent when you're making one module for a class as it is when you have to make an entire wall in practice. So I think that was a big thing, and then also, we learned a lot about how to write and submit research papers. There's a lot of things that we gained insight into by working on those papers with Andrew Saunders in a team, such as what sort of things do you include? How do you write this in the most engaging or informative way? How do you frame it? In a certain way, and all that sort of thing.

**RSS**: *All of you, do you think RAS's design and technology-based education has been empowering for you? Do you see it being instrumental to your current and future career*?

**Dan:** Yes, automation tools such as Python and Grasshopper are used almost daily here at Zahner. Having some prior knowledge about these before I started working here was a huge help. The speed with which we adopt robotic processes in the MSD-RAS program came in handy for me, it helped me secure a full-time job before my initial internship period was even over.

**Claire:** RAS allowed me to have another set of tools to work with in my design practice and taught me the process of acquiring other similar tools in the future.

**Mahsa:** I believe it's important to experience academia before entering into industry to benefit from its freedom and open-mindedness and to be trained in state-of-the-art methods that one can then bring to industry problems.

**Riley:** Absolutely. With my B.Arch, my designs were as excellent as anyone else's. With my MSD-RAS degree my designs are innovative & sexy. My design voice was amplified by a magnitude through my MSD-RAS education.

**RSS**: ***What do you hope to be doing five years from now?***

**Claire:** I hope to be teaching and making independent work in a self-sustaining practice.

**Riley:** Continuing to teach full-time, approaching or on tenure track at a leading US university,

*Opposite: Close-up of EPS foamrobot hot-wire cut volumetric study, Fall Material Agencies Robotics & Design Lab I course, students: Claire Moriarty, Matthew White.*

and continuing to employ a small team in private practice producing luxury building materials with better profit margins than now.

**Dan:** I hope to continue working in the field of construction automation.

**Mahsa:** I see myself engaging more in industrial-scale projects while pushing forward the boundaries of robotic design and fabrication, in industry or academia. I hope to be rethinking labor practices whilst expanding aesthetic possibilities in architecture. I also hope that I will be poised to establish a laboratory, disseminate knowledge globally, and transcend conventional boundaries in design and manufacturing. It's been a while since I dreamed about having my own furniture and fashion design studio, so I am working towards that.

**RSS**: *What would you say to a student considering whether or not to study in the MSD-RAS?*

**Dan:** If you are open to a fast-paced environment or trying new things with exciting tools–I would say go for it! There will be integration with architectural design ideas, scientific research and writing, fabrication, tool-making, automation, and more! The RAS faculty is also open to student feedback—this was very valuable to me. The program is relatively new and there are sufficient opportunities for students to give their opinion on how to mold it for the future! There are opportunities for TA or RA positions with faculty within and outside of the RAS program, as well.

**Mahsa:** This program will give you one year of hands-on experience working with robots and engaging in computational design, software programming, and design. These experiences will challenge your ideas and aspirations for the future of architectural design.

**RSS:** *Great, that's probably enough questions from me. But do any of you have any other observations or thoughts about the program that you would like to share?*

**Dan:** *Yeah, I think that I, you know in undergrad, I probably had a hard time making free time for myself. But there was something about after Covid, and being in grad school, that this became so much more important to me. Although being at a new school with a lot of interesting people and faculty around, somehow I found a way to manage my calendar with the RAS work and still make time to be social and do other things and meet other interesting people at Penn. So if there are students who are hesitating to sign up for a grad program, It's definitely possible to make time and do those additional social, health-orientated, or academic things while you are studying in the RAS program. I was also working part-time in a RAS robotics lab operator position which was also doable. And, you know, another point: there's another thing I liked about the RAS program, we were treated as adults, while still being allowed to be students. I really liked that, and, also, I think I was told the faculty would be amazing, and then the faculty was amazing! So I think that it would could be cool to mention this. That kind of might be too much, but that was something I liked about the RAS program.*

**RSS:** *That's a valuable comment Dan. I think these questions are on every student's mind right now. This work-life-balance issue is only growing, so I think it's good that you that you had a positive experience in the RAS program in this respect, and with the faculty! We will need to conclude this here. Thank you all for participating in this discussion. I think you know your experiences speak volumes about the opportunities the program offers, for a student studying in it, and also what it can mean for one's career afterwards. I'm really happy to see you all doing so well. Congratulations on your well-deserved success, and stay in touch!*

# EXTRA-CURRICULAR

*Faculty Research Projects*

# DEEP RELIEF

Deep Relief combines generative artificial neural network computational methods with industrial robotic manufacturing procedures to explore the potential of designing, fabricating, and expressing novel part-to-whole relationships at an architectural scale (Saunders et al. 2022a, 2022b). More specifically, the relief tests the capacity of convolutional neural networks (CNNs) to generate a language of 3D geometry to drive tool-specific robotic fabrication at the intersection of painterly representation and non-authorial composition. While artificial neural networks are not as quick or agile as biological ones, heavy brute-force cloud computing allows them to continually roll over information and slowly "learn" through non-parametric algorithms. By referencing the main features of ruled surface geometry, rails, and rulings as image features, a VGG-19 CNN can redistribute pre-embedded construction logic indicative of robotic hot-wire fabrication routines in new and unanticipated hybridized feature sets. Unimagined, self-similar ruled surface feature clusters can then be cut from blocks of EPS foam by an ABB IRB4600 6-axis robot carrying a large-scale custom hot-wire cutter end effector. Deep Relief navigates real-world constraints while simultaneously operating within both the rigorous geometric construction logic of ruled surfaces and a unique twenty-first century meronomy enabled by artificial neural networks.

— Andrew Saunders

## Credits:

University of Pennsylvania, Stuart Weitzman School of Design

***Research Lead:*** Andrew Saunders
***Robotic Research Assistants:*** Penn Praxis Fellow Riley Studebaker, Penn Praxis Fellow Mathew White, and Claire Eileen Moriarty
***Research Assistants:*** Penn Praxis Fellow Caleb Ehly, Penn Praxis Fellow Benjamin Hergert, Yujie Li, Jesse Allen, Macarena De La Piedra, and Cecily Nishimura
***Middletown Free Library Architects:*** Erdy McHenry Architecture
M***aterials and Construction:*** Universal Foam, L.J. Paollela Construction

## References

Saunders, A., R. Studebaker, and C. Eileen Moriarty. "Deep Relief." In *Traits of Postdigital Neobaroque: Pre-Proceedings* (PDNB), edited by M. Colletti and L. Winterberg, page numbers. Innsbruck: Universitat Innsbruck, 2022b.

Saunders, A., R. Studebaker, and C. Eileen Moriarty. "Deep Relief." In *Hybrids and Haecceities: Proceedings of the 42nd Annual Conference of the Association for Computer Aided Design in Architecture*, edited by M. Akbarzadeh, D. Aviv, H. Jamelle, and R. Stuart-Smith, page numbers. La Vergne: IngramSpark, 2022b.

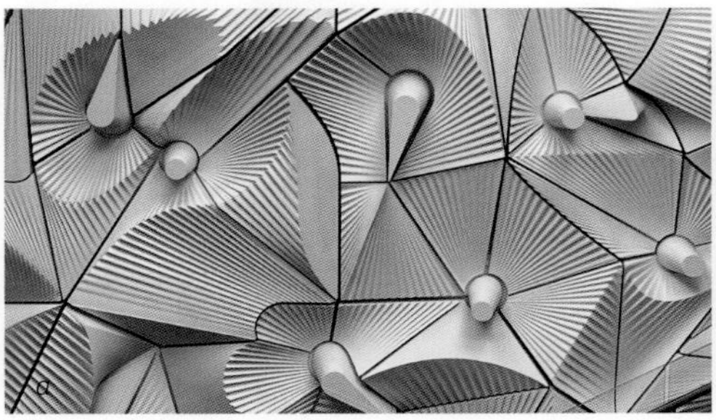

a

b

*Opposite) Photograph of final Deep Relief feature wall at Middletown Free Library, a) close-up photograph of the feature wall, and b) images from convolutional neural network (CNN) AI design process that was trained on robot hot-wire cut geometrical prototypes.*

*a*

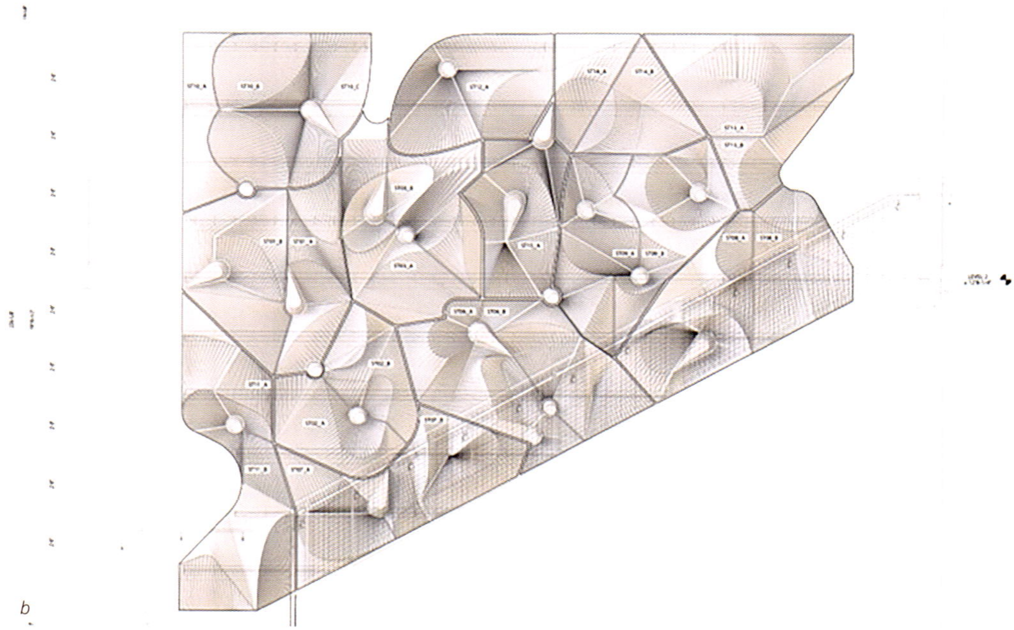

*b*

*a) Exploded elevation indicating each individual robot hot-wire cut element and an b) assembled elevation indicating how multiple hot-wire cut elements are pre-assembled into larger components prior to on-site installation.*

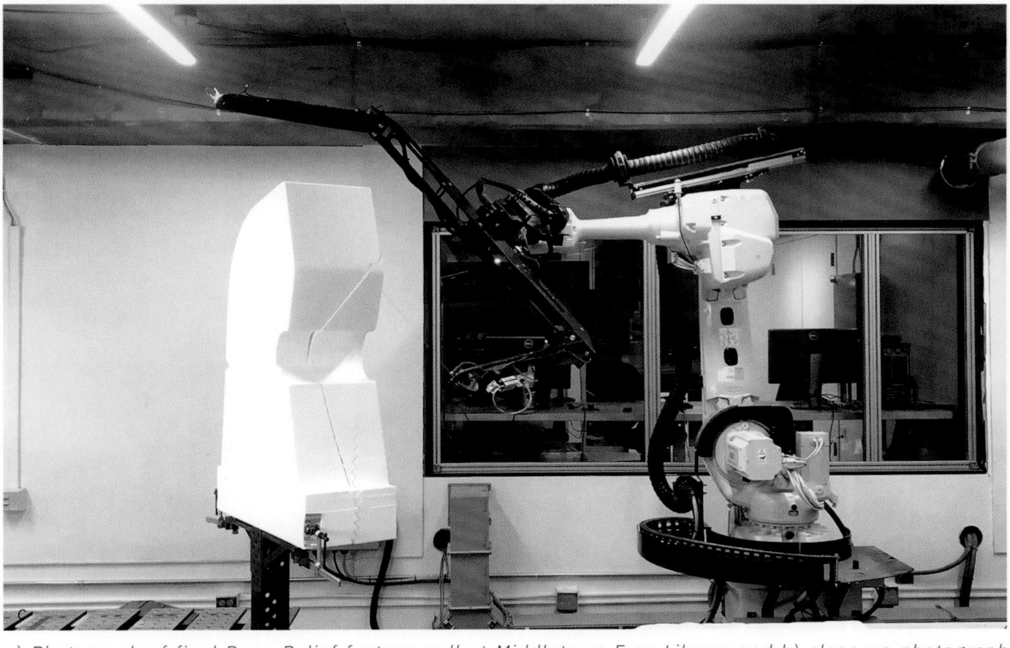

a) Photograph of final Deep Relief feature wall at Middletown Free Library and b) close-up photograph of the feature wall. c) EPS foam for the relief is cut by robot hot-wire cutting in Penn's Weitzman School of Design ARI Robotics Lab where MSD-RAS students also undertake their robotic fabrication activities.

# SITELESS HOUSE

Siteless House explores an alternative approach to relocatable housing through the development of a new type of Design-For-Disassembly (DFD) lightweight, small-scale, precast concrete tectonic that operates as a structural building envelope with integrated floor and foundations. A design-computation and fabrication method for robot hot-wire cutting of expanded polystyrene (EPS foam) was developed to enable the design and engineering of lightweight, high-strength loose-fiber reinforced precast concrete building component parts. Each part can be produced with variable geometrical form and surface articulation in order to further increase the material efficiency of each part, while exploring scale-free modes of architectural design expression. The design leverages other integrated robotic fabrication and computational methods for other parts of the house including a thermoplastic additively manufactured bathroom and an incrementally formed kitchen bench. Siteless House is a collaboration between the research labs of Weitzman Faculty Robert Stuart-Smith and Masoud Akbarzadeh together with Cemex Global R&D. The project is funded by Cemex Global R&D and Penn Alumni Hanley Bodek and involves several MSD-RAS graduates and current students. A full-scale prototype of the house will be constructed on Penn's campus soon.

a

b

## Credits:

*University of Pennsylvania Stuart Weitzman School of Design & Cemex Global R&D*

***Penn leads:*** Robert Stuart-Smith and Masoud Akbarzadeh

***Cemex Global R&D Lead:*** Vice President Davide Zampini

***Penn team:*** Bowen Qin, Mahsa Masalegoo, Claudia Campuzano, Riley Studebaker, Mariana Righi, David Forero, Yi Dazhong, and Sarah Borders.

***Cemex Global R&D team:*** Alexandre Guerini, Carlos Enrique Terrado, Valentina Rizzo, and Matthew Meyers.

c

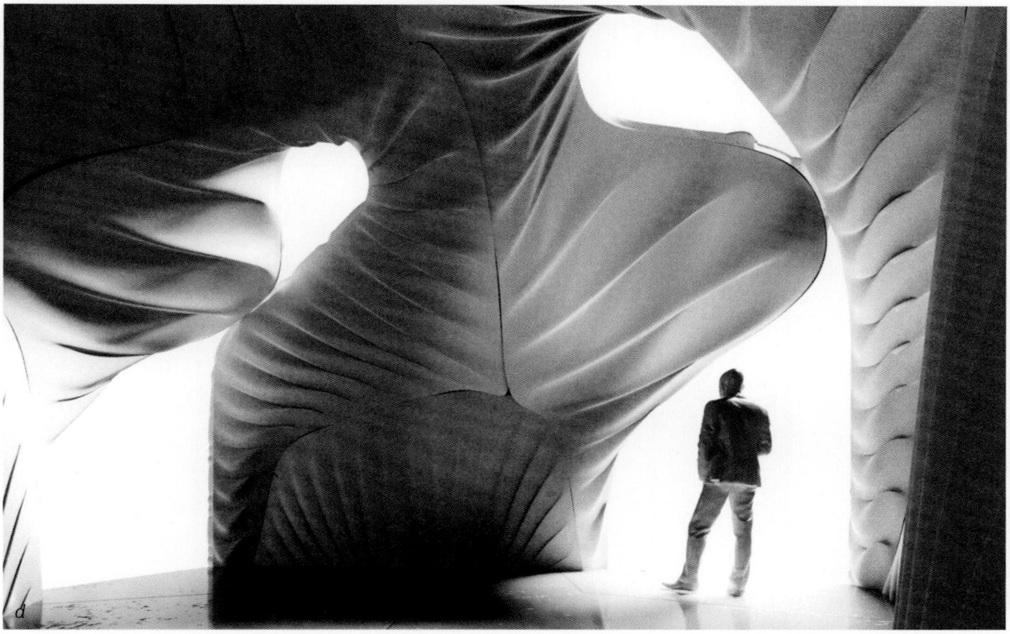

d

*a) Robot hot-wire-cut EPS foam formwork's profiling adds rigidity to precast concrete panel, b) robot hot-wire cutting process, c) siteless house exterior showing small, lightweight and unique precast concrete panels, and d) felt fabric pleated interior that visually reinforces the ruled-surface geometry of the house – a constraint of the wire-cutting process.*

# TOWARDS AN AERIAL AM ARCHITECTURE

The construction industry is adopting additive manufacturing (AM) technologies for onsite construction due to their ability to reduce the time and cost of building. On-site AM typically involves the continuous extrusion of horizontal layers of material using a gantry larger than the build volume, to move a single extruder mechanism. Transporting and installing such gantry systems limits AM to easily accessible sites. In contrast, Aerial Additive Manufacturing (Aerial AM) enables swarm-based, parallel manufacturing in remote or hard-to-access locations, providing greater flexibility and boundless manufacturing. Having recently published in *Nature* a demonstration of the world's first Aerial AM with cementitious and composite materials in-flight (Zhang et al. 2022), ongoing research includes enhancements to the manufacturing approach and a series of architectural design demonstrations of the technology through simulation and/or physical manufactured outcomes. Aerial AM research was recently exhibited at the 2023 Venice Biennale of Architecture (Stuart-Smith, Kovac, et al. 2023), several videos of Aerial AM technology were featured adjacent to architectural designs developed using the same robot control framework in simulation to speculate on the potential of aerial additively manufactured architecture. Two projects were presented: a tower of near-infinite height (Ander et al. 2019) and a large-span shell (Stuart-Smith, Darekar, et al. 2023). Both explore the logistical and aesthetic possibilities of an unbounded swarm-constructed architecture that is partially designed through the act of aerial swarm construction.

## Credits:

***Institutions:*** University of Pennsylvania, Imperial College London, University College London, University of Bath, and Empa.

***Aerial AM Venice Biennale Exhibition:*** Robert Stuart-Smith, Mirko Kovac, Franklin (Renhu) Wu, Yusuf Furkan Kaya, Lachlan Orr, and Adam Blood.

***Aerial AM Shell:*** Robert Stuart-Smith, Patrick Danahy, Mirko Kovac, and Vijay Pawar.

***Aerial AM Tower:*** Robert Stuart-Smith, Chris Williams, Paul Shepherd, Vijay Pawar, Mirko Kovac, Andrew Homick, and Patrick Danahy.

***Aerial AM Project leaders:*** Mirko Kovac, Robert Stuart-Smith, Stefan Leutenegger, Vijay Pawar, Chris Williams, Paul Shepherd, and Richard Ball. For more information, See Zhang et al., *Nature*, 609, no. 7928 (2022): 709-717.

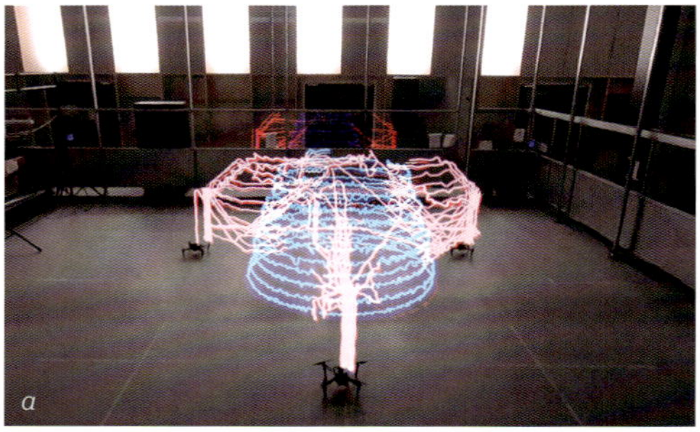

*a) Aerial AM light-trace time-lapse made by several UAVs virtually printing (Photo by University College London, Imperial College London, University of Bath. First published in* Nature *(Zhang et al. 2022)). b) Venice Architecture Biennale 2023 exhibit.*

*a) Aerial AM demonstration, 3D printing in flight with custom cementitious material (Photo by Imperial College London, University College London, University of Bath. First published in* Nature *(Zhang et al. 2022)). b) Closeup of Aerial AM Tower exoskeleton and c) closeup of Aerial AM Shell project.*

## References


Ander, M., P. Shepard, R.Stuart-Smith, and C.Williams. "A building of unlimited height". In *Proceedings of the IASS Annual Symposium 2019 - Structural Membranes 2019: Form and Force*, edited by C. Lázaro, K.-U. Bletzinger, and E. Onate (Eds.), International Association for Shell and Spatial Structures (IASS), Barcelona, 2019, 1426-1433

Stuart-Smith, R., D. Darekar, P. Danahy, B. Bahadir Kocer, V. Pawar, and M. Kovac. "Collective Aerial Additive Manufacturing." In *Proceedings of the 42nd Annual Conference of the Association of Computer Aided Design in Architecture (ACADIÁ)*, edited by M. Akbarzadeh, D. Aviv, H. Jamelle, and R. Stuart-Smith, 44-55. La Vergne: IngramSpark, 2023.

Stuart-Smith, R., M. Kovac, V. Pawar, C. Williams, and P. Shepard. "Aerial Additive Manufacturing" (Aerial AM). In *Time, Space, Existence: Venice 2023 Architecture Biennial*, edited by European Cultural Centre Team, 474-475. Amsterdam: European Cultural Centre, 2023.

Zhang, K., P. Chermprayong, F. Xiao, D. Tzoumanikas, B. Dams, S. Kay, B.B. Kocer, A. Burns, L. Orr, T. Alhinai, C. Choi, D.D. Darekar, W. Li, S. Hirschmann, V. Soana, S.A. Ngah, C. Grillot, S. Sareh, A. Choubey, R. Stuart-Smith and M. Kovac, et al. "Aerial Additive Manufacturing with Multiple Autonomous Robots." *Nature* 609, no. 7928, 2022: 709-717. DOI. https://doi.org/10.1038/s41586-022-04988-4.

# EPILOGUE

*A Brief Conclusion*

# MSD-RAS: SHAPING TOMORROW TODAY

*Robotic Fabrication and Architectural Design* is the first book on the University of Pennsylvania's MSD-RAS program. The book presented a series of robotic fabrication and material production approaches that were explored creatively and discursively during the first three years of the program. Operating from the Department of Architecture in the Stuart Weitzman School of Design, the program's ambitious and technically progressive curricula are engaged through acts of design. Experimental prototyping becomes the means by which several different fields of knowledge are synthesized in a series of creative projects that investigate different approaches to robotic manufacturing and computational design. The curricula are intended to provide students with an intensive and effective learning experience whilst also offering room for students to conduct exploratory research to develop their own final semester thesis project.

The book showcased the design and production activities of students working across several different project briefs. This work was supported by essays that described what makes the MSD-RAS program unique while course summaries conveyed some of the different ideas and themes explored in the curricula. The book also offered insights into the experiences of the program's graduates through interviews and survey data, highlighting the diverse and successful careers that have emerged during these initial years. It is hoped that this confluence of the program's work and philosophy, combined with the experiences of its graduates, will inspire some readers to apply for admission to the MSD-RAS. And for others, encouragement to seek out more diverse and technically creative agency in industry—in areas presently considered to be beyond the scope of architectural practice.

At the time this book is being published, the MSD-RAS program is launching into its fifth year, engaged in entirely different material, fabrication, and research themes that will likely be the subject of a future follow-on book. The program continues to question established architectural design and industrial manufacturing processes, developing alternative design methodologies that integrate materials, manufacturing techniques, and computational approaches to architectural design and robotic fabrication. We hope this book serves as both a point of reflection and a source of inspiration, illustrating how design research can enhance and expand our current manufacturing capabilities to support more ecologically minded and distinctively crafted possibilities for tomorrow's built environment.

– Robert Stuart-Smith, MSD-RAS Program Director.

*Close-up of a glazed ceramic 3D printed façade screen prototype part. Viewed as an artefact, the part is embodied with specific design characteristics, developed in the design research. These qualitative, aesthetic aspects of the part are intrinsic to the material and custom-developed manufacturing method (Robotic Prometheus project, spring semester).*

# APPENDIX

*Faculty Bios*

## Robert Stuart-Smith

MSD-RAS Program Director

Robert Stuart-Smith is director of the MSD-RAS degree program, assistant professor of Architecture, and affiliate faculty in Engineering's GRASP Lab at the University of Pennsylvania. He directs the Autonomous Manufacturing Lab in both Penn's Department of Architecture and University College London's Department of Computer Science, leading over $5 million in government-funded research, collaborating with industry partners Cemex, Skanska, Mace, Burohappold, Arup, MTC, Ultimaker, and others. Stuart-Smith's research operates at the intersection of design, computation, robotic fabrication, and collective robotic construction. His multidisciplinary Aerial Additive Manufacturing research published in *Nature*, demonstrates the world's first additive manufacturing in-flight, undertaken by several cooperating drones. By co-developing robotic manufacturing capabilities alongside integral architectural design methods, Stuart-Smith's work aims to address the environmental and economic cost of design and production together with their unified potential to offer a cultural and aesthetic impact. Stuart-Smith co-directs the architectural practice Robert Stuart-Smith Design and previously co-founded the experimental design practice Kokkugia in 2004 after working at Arup, Grimshaw and Lab Architecture Studio, altogether working on 1.6mil SqFt and $1.7bil of built work. Author of Routledge's book *Behavioural Production*, his work has also been published in journals including *Nature, Science Robotics, AD Architectural Design,* and *Architecture D'Aujourd'hui*. Included in the permanent collection at Frac Centre-Val de Loire, his work was also exhibited in Venice, Tallinn, and Prague Architecture Biennales, and Paris Lieu de Design. Stuart-Smith previously taught at AA, WashU, RMIT, and U.Innsbruck. He has lectured at ETHZ, U.Stuttgart, MIT, CCA, Sci-Arc, AA, UCL, Angewandte, Strelka Institute, Tsinghua University, and been interviewed by BBC Click, New Scientist, Smithsonian, Architizer, and others.

## Dr. Nathan King

Dr. Nathan King is a senior industry futures manager at Autodesk and contributes to strategic initiatives related to robotics, automation, manufacturing, and construction. In addition, he was instrumental in the creation of the Autodesk Technology Centers and continues to work with industry to catalyze innovation opportunities. Prior to joining Autodesk, Nathan was a director at Mass Design Group where he focused on the development of building technologies, medical devices, and evaluation methods for use in resource-limited settings. King is active in industry, practice, academia, and policy—working to create opportunities at their intersection. Nathan's work spans the globe, with built work in Malawi, Rwanda, Tanzania, Uganda, Haiti, Sweden, Dominican Republic, United States, and beyond. King is co-diretor of the Center for Design Research at Virginia Tech where he leads several programs focusing on design technology and impact design, including an initiative focused on the democratization of design technology through collaboration with the regions HBCUs and funded research focusing on automation in construction in the US's rural south. He is also a lecturer in Architecture at the Harvard University Graduate School of Design where he teaches courses in digital material systems and instructor at the University of Pennsylvania Stuart Weitzman School of Design, focusing on robotics and autonomous systems in design and construction. Nathan regularly teaches as part of AIA, ACADIA , RobArch, and other conferences, lectures widely, and publishes often including co-authoring of the book Ceramics Material Systems in Architecture and Interior Design. Currently, Dr. King is co-developing a Liberian vocational school that will close skills gaps and provide job opportunities in the country's agriculture, textiles, and construction trades.

## Billie Faircloth

Billie Faircloth, FAIA, is an architect, educator, and transdisciplinarian who has transformed practice-integrated research and earned a national and international reputation for demonstrating its value, methods, and outcomes. At Weitzman School of Design, Billie is an adjunct professor of Architecture, teaching in two MSD programs, Robotics and Autonomous Systems (RAS), and Environmental Building Design (EBD). She formerly taught at The University of Texas at Austin and Harvard GSD. She served as the Barber McMurry Visiting Professor of Practice at the University of Tenessee Knoxville and VELUX Visiting Professor at the Royal Danish Academy. As a former partner and research director at the firm KieranTimberlake from 2008 to 2024, Billie worked to expand the problem-solving boundary for professionals, fostering collaboration between trades, academies, and industries. Billie led the firm's research group, a team leveraging methods from fields as diverse as urban ecology, materials science, chemical physics, and architecture. Billie's co-authored projects and published studies address knowledge gaps and contribute to building technology development. Her projects include TallyLCA, Roast, Ideal Choice Homes, and Multi-scale Modeling. Her articles and book chapters are published by the *International Journal of Architectural Computing, Journal of Architectural Education, Princeton Architectural Press*, and *SimAUD*. *Architectural Record* celebrated Billie's pioneering practice-based research by selecting her as its 2017 Women in Architecture Innovator Award winner. Billie has served the American Institute of Architects, Committee on the Environment (AIA COTE), and the International Union of Architects (UIA) World Congress of Architects. Billie joins Cornell University in 2025 as an associate professor in the Department of Architecture and a Senior Faculty Fellow at the Cornell Atkinson Center for Sustainability.

## Andrew Saunders

Andrew Saunders is an associate professor of Architecture at the University of Pennsylvania Stuart Weitzman School of Design where he serves as the associate chair of the Department of Architecture and director of the Master of Architecture Professional Degree Program, as well as teaching studios in the Master of Science in Design: Robotics and Autonomous Systems (MSD-RAS) program. He is founding principal of Andrew Saunders Architecture + Design, an internationally published, award winning architecture, design and research practice committed to the tailoring of innovative digital methodologies to provoke novel exchange and reassessment of the broader cultural context. The practice innovates at several scales ranging from product design, exhibition design, and residential and large-scale civic and cultural institutional design. He received his Bachelor of Architecture from the University of Arkansas and a masters in Architecture with Distinction from the Harvard Graduate School of Design. His current practice and research interests lie in computational geometry as it relates to aesthetics, emerging technology, fabrication, and performance. Due to the multiplicity of overlapping research interests, he has published significantly in an array of fields. Recent work includes groundbreaking LiDAR survey analysis in his book, *Baroque Topologies* published by Palombi Editori and unprecedented integration of robotic fabrication with AI design methodologies in the permanent installation, Deep Relief. He has significant professional experience as project designer for Eisenman Architects, Leeser Architecture, and Preston Scott Cohen, Inc. He has taught and guest lectured at a variety of institutions, including Cooper Union and the Cranbrook Academy of Art, and was an assistant professor of Architecture and head of graduate studies at Rensselaer Polytechnic Institute in New York.

## Jeffrey Anderson

Jeffrey Anderson is an educator, architectural designer, and AR/VR software developer. He currently teaches design studios and advanced media seminars in the Graduate Architecture and Urban Design program at Pratt Institute and the Graduate Architecture Program at the University of Pennsylvania. He is also the lead software developer in the Design Lab at Mancini Duffy where he conducts design research and develops architectural visualization tools. His current software development work focuses on creating new forms of physical and virtual collaboration that empower all members of the design process. His research focuses on using technology to create new relationships between users, architecture, and its context through interaction, sensing and feedback, and mixed reality. His recent book, *The Ecologies of the Building Envelope: A Material History and Theory of Architectural Surfaces* (Actar 2021), written with Alejandro Zaera Polo, analyses how social, political, technological, and economic forces have become embedded within architecture over the last century. His 2017 book, *Imminent Commons: The Expanded City*, edited with Alejandro Zaera Polo situates contemporary urbanism through the lens of ecological and technological resources relevant to all cities. He holds a Master of Architecture II from Princeton University, and both a Master of Architecture and a Bachelor of Science in Architecture from the Knowlton School of Architecture at Ohio State University.

## Ezio Blasetti

Ezio Blasetti, registered architect in Europe (TEE-TCG), holds a Masters of Science in Advanced Architectural Design from Columbia University after having previously studied in Athens and Paris. His academic and professional research focuses on the application of advanced technologies in all phases of architectural design, from the initial composition to the digital and robotic fabrication. He is a founding partner at Maeta Design, an architectural design and research firm based in Philadelphia, New York, and Athens. In 2009, he co-founded Ahylo, an architectural design and construction practice and "Apomechanes", an intensive design lab on al gorithmic processes and fabrication. Founder of algorithmic design.net,Blasetti's recent collaborations include New Territories, Biothing, Acconci studio, a|Um studio, and serge studio. He has taught generative design studios and seminars by means of computational geometry and digital fabrication at Weitzman School of Design, Columbia University, Yale SoA, Pratt Institute, Rensselaer Polytechnic Institute, Sciarc, Cooper Union, the Architectural Association, Michigan University, University of Technology Sydney, and the Bartlett. His work has been exhibited and published internationally and is part of the permanent collection of the Centre Pompidou.

## Evangelos Kotsioris

Evangelos Kotsioris is assistant curator in the Department of Architecture and Design at the Museum of Modern Art in New York. Among other projects, he has recently co-organized the exhibitions "Reuse, Renew, Recycle: Recent Architecture from China" (2021–22) and "The Project of Independence: Architectures of Decolonization in South Asia, 1947–1985" (2022). Prior to MoMA, he was the 2016–17 Emerging Curator at the Canadian Centre for Architecture in Montreal, where he presented the exhibition "Lab Cult: An Unorthodox History of Interchanges Between Science and Architecture" (2018). He has worked as an architect at OMA/AMO in Rotterdam and holds an Masters of Architecture II from Harvard GSD. His doctoral research on the history of architectural computerization in the Soviet Union at Princeton University was awarded the Carter Manny Citation for Special Recognition by the Graham Foundation in Chicago. Kotsioris has taught at Barnard+Columbia Colleges, Penn, Princeton SoA, The Cooper Union, and Harvard GSD. His writing has appeared in *Perspecta, New Geographies, The Architectural Review, Volume, Manifest, post* (at MoMA), and elsewhere. He is co-editor of *Radical Pedagogies*, a global history of experiments in architectural education (MIT Press 2022).

## Alicia Nahmad Vazquez

Alicia Nahmad is an architect with a passion for robotics and digital fabrication. As a research-based practicing architect for the last 12 years, she has been engaged with the digitization of building trades and adapting advanced digital design and robotic fabrication methods to incorporate the wisdom and craft of traditional building cultures. Her projects include the construction of award-winning Knit-Candela and diverse collaborations with practice and academic institutions such as ZHA CODE,Block research group at ETHZ, AAC, ODICO, and more. She holds a PhD from Cardiff University and a Masters of Architecture from the AADRL. She has developed workflows for human-robot collaboration in the design and construction process that engage with local communities using advanced technologies. Alicia's work expands across a number of scales, from architectural pavilions to robotic installations and furniture. As an academic and an entrepreneur, Alicia is the founder of The Circular Factory (CF) and MITollbox. Alicia works as an associate professor at the University of Calgary SAPL. She also co-directs the Laboratory for Integrative Design. Before joining SAPL, she worked as studio master at the AADRL for 5 years. Previously, Alicia worked developing digital tools for practices like Populous and Zaha Hadid Architects. Her work has been published in numerous conferences and journals an she has lectured internationally on related topics.

## Jose Luis Garcia del Castillo Lopez

Jose Luis García del Castillo y López is an architect, computational designer, and educator. He advocates for a future where programming and code are tools as natural to artists as paper and pencil. In his work, he explores creative opportunities at the intersection of design, technology, fabrication, data, and art. García del Castillo y López is a registered architect, and holds a Doctor of Design, and Master in Design Studies in Technology from Harvard University Graduate School of Design. As a structural engineer at Ayesa, he has consulted for international firms such as OMA, Mecanoo, and Cesar Pelli. Additionally, he has worked as a data visualization architect at Fathom Information Design, designing and building platforms to help clients such as Google, Samsung, and Nike understand their data. More recently, he has worked as a Research Engineer within the Generative Design Group at Autodesk Inc., and lectured in the MSD-RAS program at UPenn. García del Castillo y López is also the co-founder of ParametricCamp, an international organization whose mission is to spread the knowledge of computational design among designers and architects. His current research focuses on the development of digital frameworks that help democratize access to robotic technologies for designers and artists. In his doctoral dissertation, he proposed Enactive Robotics, a new model for enhanced human-robot interaction using machines as an exploratory platform through concurrent control. When applied to digital fabrication, for example, this model systematically enables closed-loop control mechanisms, opening up the possibility of incorporating material feedback or human mediation in real-time.

## Emek Erdolu

Emek Erdolu is a PhD. candidate in Computational Design at Carnegie Mellon University (CMU) School of Architecture. Situated across architecture, design computation, human-computer interaction, and robotics, his research includes two main strands: one that focuses on building intuitive and interactive computational tools to support architectural practices, research, and education; and another one on the questions of computation, representation, and work within historical and contemporary architectural practices. His PhD research focuses on AI/robotic systems we work with every day. His forthcoming dissertation investigates how building tasks are analyzed and decomposed to inform on-site robotic systems and how these systems reconfigure the organization of these tasks with new workflows and human-machine interactions they introduce to construction sites, focusing on a selection of historical and contemporary robotics practices in the United States. Part of this research and more have also been co/published in the Proceedings of the ACM on Human-Computer Interaction (PACMHCI) and International Journal of Architectural Computing (IJAC),and presented in venues such as the Society for Social Studies of Science (4S). Previously, Emek was a researcher at the ETH Future Cities Laboratory (FCL), and for seven years in the United States, China, and Southeast Asia, he worked in various architecture, landscape architecture, and urban design projects with AECOM, HMD, Ecoland, and Nomad Studio. He has lectured and taught studios in CMU, Singapore University of Technology and Design (SUTD), National University of Singapore (NUS), and Bilkent University. Aside from MSD-RAS at Weitzman, he currently teaches at CMU's Master of Science in Computational Design (MSCD) program.

## Patrick Danahy

Patrick Danahy is an assistant research professor of Architecture and the Design Innovation Fellow at Ball State University and the 2022 distinguished professor of the T4T Lab at Texas A&M, a position formerly held by Casey Rehm, Roland Snooks, Tom Wiscombe, Barry Wark, Gilles Retsin, Nate Hume, and other distinguished faculty. He has taught digital workshops in the UCL Bartlett RC20 Skills-Share program, as well as design and technology studios at the University of Pennsylvania as a part-time lecturer for the Master of Architecture and MSD-RAS programs, and as a visiting lecturer for the Landscape Architecture master's program. His teaching focuses on computation and robotics, integrating architectural reference with contemporary machine learning methods. Danahy holds a Bachelor of Arts in Architecture from Clemson University, where he received the graduating Faculty award and the Peter Lee and Kenneth Russo Award for Design Excellence. He later graduated from the University of Pennsylvania with a Master's of Architecture, receiving the Kanter-Tritsch Prize in Energy and Architectural Innovation, the Paul Cret T-Square Fellowship, the Van Alen Traveling Fellowship, the Dales Fellowship, the Kohn Fellowship, the Schenk-Woodman Merit Award, and, upon graduating, was awarded the Arthur Spayd Brooke Memorial Silver Medal. His work has been presented and published at the 2020 Digital Futures Young conference, the 2020 Distributed Proximities ACADIA conference, and the 2022 CAADRIA Post Carbon conference, receiving the award for best presentation runner up with Robert Stuart-Smith.

## Teaching Assistants 2020-2022

Patrick Danahy, David Forero, Hadi El Kebib, Sophia O'Neil, Riley Studebaker, Claire Moriarty, Matthew White, Caleb Ehly, and Yujie Li.

## ARI Robotics Lab Managers

The ARI Robotics Lab facility is where MSD-RAS students undertake most of their work. Full-time lab managers' primary role is in the development, safety, and operational aspects of the facilities, and in managing several part-time staff. However, beyond the faculty and TAs, lab managers also support student activities in the robotics lab. In the MSD-RAS program, robotics lab managers are essentially faculty, knowledgable in the work and methods of the program and excited to work with students to support their project work.

## Nicholas Sideropoulos

Nicholas Sideropoulos is an Advanced Research & Innovation Lab (ARI) Robotics Lab Manager at the University of Pennsylvania, Weitzman School of Design. Sideropoulos holds a BArch from Rensselaer Polytechnic Institute and a Master of Science in Design: Advanced Architectural Design from the University of Pennsylvania's Weitzman School of Design.

## Shunta Moriuchi

Shunta Moriuchi is an Advanced Research & Innovation Lab (ARI) Robotics Lab Manager at the University of Pennsylvania, Weitzman School of Design. Moriuchi holds a BArch from Calfornia College of the Arts and a Master of Science in Design in Robotics and Autonomous Systems from the University of Pennsylvania.

## Former ARI Robotics Operators

Sophia O'Neil (2023), Riley Studebaker (2022), David Forero (2020-21).

## ARI Robotics Lab Senior Management

Robert Stuart-Smith and Karl Wellman

Published by ORO Editions,
Gordon Goff: Publisher

www.oroeditions.com
info@appliedresearchanddesign.com



Edited by Robert Stuart-Smith

Contributors: Jeffrey Anderson, Billie Faircloth, Nathan King, Andrew Saunders, Ezio Blasetti, Patrick Danahy, Jose Luis Garcia del Castillo Lopez, and Evangelos Kotsioris.

Book Design and typesetting: Robert Stuart-Smith, Yang Meng, Adam Blood.

Project Manager: Jake Anderson

Typeset in IBM Plex Sans and Maxeville Construct

First Edition

ISBN: 978-1-957183-96-1

Color Separations and Printing: ORO Editions Inc.
Printed in China.

ORO Editions makes a continuous effort to minimize the overall carbon footprint of its publications. As part of this goal, ORO, in association with Global ReLeaf, arranges to plant trees to replace those used in the manufacturing of the paper produced for its books. Global ReLeaf is an international campaign run by American Forests, one of the world's oldest nonprofit conservation organizations. Global ReLeaf is American Forests' education and action program that helps individuals, organizations, agencies, and corporations improve the local and global environment by planting and caring for trees.